Lecture Notes in Logistics

More information about this series at http://www.springer.com/series/11220

Hans-Jürgen Sebastian · Phil Kaminsky
Thomas Müller
Editors

Quantitative Approaches in Logistics and Supply Chain Management

Proceedings of the 8th Workshop on Logistics and Supply Chain Management, Berkeley, California, October 3rd and 4th, 2013

Springer

Editors
Hans-Jürgen Sebastian
RWTH Aachen University
Aachen
Germany

Thomas Müller
RWTH Aachen University
Aachen
Germany

Phil Kaminsky
IEOR Department
University of California
Berkeley, CA
USA

ISSN 2194-8917 ISSN 2194-8925 (electronic)
Lecture Notes in Logistics
ISBN 978-3-319-12855-9 ISBN 978-3-319-12856-6 (eBook)
DOI 10.1007/978-3-319-12856-6

Library of Congress Control Number: 2014956482

Springer Cham Heidelberg New York Dordrecht London

Printed on acid-free paper

Springer International Publishing AG Switzerland is part of Springer Science+Business Media (www.springer.com)

Preface

The primary goal of these workshops is to bring together people who are involved in the academic/practitioner interface in Logistics and Supply Chain Management.

The first three workshops of this series were held between 2003 and 2005 in Berkeley, CA, while the fourth workshop was in Hamburg, Germany, in 2006. Since 2007 the workshop has taken place every 2 years in Berkeley. Maintaining this tradition, the eighth workshop was held there in October 2013 at the Claremont Hotel Club & Spa.

This book is a collection of proceedings of this latest workshop. All of the papers have been written by academic researchers in the field or industry leaders who have an active consulting practice in Supply Chain Management. Each paper has been carefully selected and reviewed by members of the Workshop Committee.

Addressing *Quantitative Approaches in Logistics and Supply Chain Management*, the proceedings focus on the following topics:

- Vehicle Routing and Scheduling;
- Hub Location;
- Supply Chain Management;
- Courier, Express, and Parcel Service Network Design;
- Health Care Planning and Scheduling.

Hans-Jürgen Sebastian
Phil Kaminsky
Thomas Müller

Contents

Part I
Vehicle Routing and Scheduling

Modeling Mixed Load School Bus Routing

James F. Campbell, Jeremy W. North and William A. Ellegood

Abstract Transporting pupils to and from schools is a complex and expensive logistics problem for many public school districts, especially in rural areas where travel distances are longer. In many regions of the world, students ride public transit to school, but public school districts in the US and Canada generally provide transportation in dedicated school buses. Each bus typically makes a sequence of trips each morning and each afternoon, where each trip serves a separate school, usually with staggered start times for different school levels (elementary school, intermediate school, high school). This research explores whether the successful business logistics practice of mixed loading can be applied to school bus transportation. Mixed load school bus trips carry students for more than one school at the same time, and a mixed load routing policy reduces the number of stops to pick up and drop off students, but it adds travel distance at the end of a trip to visit multiple schools. We first providc a general strategic analysis using continuous approximation modeling to assess the conditions under which mixed loading is likely to be beneficial. Then we present a discrete algorithm for finding mixed load bus trips. Results for benchmark data sets explore the tradeoffs between minimizing the number of buses used and minimizing the travel distance. We also present a case study for a Missouri school district to illustrate the application of the models in practice. Results show that mixed load bus routing can be beneficial when students are sparsely distributed, when a large percentage of bus stops are shared by students of different schools, and when schools are closer together.

J.F. Campbell (✉)
College of Business Administration, University of Missouri – St. Louis,
One University Blvd., St. Louis, MO 63121-4499, USA
e-mail: campbell@umsl.edu

J.W. North
College of Business, Murray State University,
4430 Sunset Avenue, Paducah, KY 42001, USA
e-mail: jnorth@murraystate.edu

W.A. Ellegood
College of Business Administration, Sam Houston State University,
1821 Avenue I, Box 2056, Huntsville, TX 77341, USA
e-mail: wellegood@shsu.edu

H.-J. Sebastian et al. (eds.), *Quantitative Approaches in Logistics and Supply Chain Management*, Lecture Notes in Logistics,
DOI 10.1007/978-3-319-12856-6_1

1 Introduction

An important logistics problem for public school districts is the transportation of pupils to and from schools in a safe, reliable, and cost effective manner. While many regions around the world rely heavily on students using existing public transportation systems, in the US and Canada, public school districts provide transportation in dedicated school buses for all students who live more than a specified distance from their school. According to the American School Bus Council, school bus transportation represents the largest form of mass transit in the US, with over 480,000 school buses transporting about 26 million children each day, which is over half of the schoolchildren in the US [2]. Further, they estimate that US school buses travel 5.76 billion miles per year at a cost of $21.5 billion, while providing a much more efficient and environmentally friendly alternative to private automobiles.

School bus transportation may be provided by a school district using its own private fleet of buses, or by a third party under contract to the school district. Recently, increasing fuel prices and the economic downturn have increased pressures on optimizing school bus transportation. Higher fuel prices translate to increased costs for school bus transportation and have led many districts to redesign routes to reduce travel distance and/or cut the availability of buses to some students [55]. The recent economic recession has eroded tax revenues and increased pressures on regional and local governments that typically provide funding for public school bus transportation.

As a concrete example, in Missouri (an "average" US state in terms of population and area), school buses travelled nearly 124 million miles transporting more than one-half million students to and from schools daily [42]. Missouri state statutes require public school districts to make transportation available for all students that live more than 3.5 miles from their school, but the state provides reimbursement of a significant portion of transportation expenses for all students that live more than 1 mile from school. Consequently, most Missouri school districts make transportation available for all students who reside more than one mile from school, and some even use a smaller distance of 0.5 mile. However, as a result of the recent economic decline, Missouri reduced state aid for pupil transportation by 46 % [34], which forced individual school districts to consider changes in school bus transportation plans and budget cuts in other educational areas to offset the decreased support from the state.

Each local school district develops its school bus routing plans, with modifications each year to reflect changes in enrollments, school building operations, transportation costs, etc. Many districts still design school bus routes with a largely manual process, though routing software packages and services provided by contractors are used in some districts. However, the complexities of school bus routing (e.g., safety and security restrictions, multiple routes per bus for different schools, varying start times, uncertainties regarding the number of riders each day, variable loading and unloading times, etc.) and idiosyncratic local conditions (traffic, weather, operating policies and traditions) make many software solutions impractical without considerable manual input [40, 60].

This research extends the business logistics concept of mixed loads (e.g., carrying a blend of different products on a single vehicle) to school bus routing. We consider a typical morning trip of a school bus that carries students from many bus stops to their school(s). A *non-mixed bus trip* transports students bound for a single school from many bus stops to that school. A *mixed bus trip* transports students bound for two or more schools to those schools. For example, a mixed trip for two schools will pick up students for both schools at each bus stop and then deliver them in turn to the two schools. Thus, a mixed bus trip carries students for two (or more) different schools at the same time, while a non-mixed bus trip carries students for only a single school. A typical bus routing plan uses a sequence of non-mixed bus trips, each serving a single school, with all schools at each level (e.g., elementary schools, middle schools, high schools) having the same starting (and ending) time. Some school districts have employed mixed loads in places, but it is not widespread and a general analysis of mixed loading is needed.

Our primary objectives are (1) to explore the utility of mixed load school bus trips using a strategic analytical model for a generic school district, and (2) to develop a discrete algorithm for finding good mixed load school bus trips, with a focus on minimizing the total bus travel distance. We first undertake a general strategic analysis using continuous approximation models for school bus transportation in a generic school district to estimate potential savings from mixed trips. Analytical continuous approximation models are based on a continuous spatial density of demand, rather than discrete locations of demand points and they are useful for strategic and policy analysis in transportation [19]. The strategic analysis allows an assessment whether more detailed analysis would be worthwhile in light of potential savings in bus travel distance and the expected disadvantages from mixed trips. However, the strategic continuous approximation models do not provide actual bus routes for a particular setting.

The second area of analysis is development of a discrete algorithm to determine mixed load bus trips given a set of bus stops, schools and students to be transported. This complements the strategic continuous approximation models with a composite heuristic routing and scheduling algorithm that generates mixed load bus routes. This algorithm is tested on benchmark data sets with up to 2,000 bus stops and 100 schools and results are compared to an alternative approach in the literature [48]. Finally, to tie together the two approaches we provide a case study for a semi-rural school district in eastern Missouri to test both the utility of the analytical continuous approximation models and the discrete algorithm in light of real-world complexities.

The remainder of this paper is organized as follows. Section 2 reviews the relevant literature on both continuous approximation modeling and discrete school bus routing research. Section 3 develops continuous approximation models for mixed and non-mixed school bus trips in a generic school district, and compares mixed and non-mixed routing under a variety of conditions. Section 4 presents the discrete mixed load bus routing algorithm, along with results for benchmark data sets. Section 5 is the case study for a Missouri school district and Sect. 6 is the conclusion and suggestions for future research.

2 Literature Review

School bus routing is a particularly challenging variant of the vehicle routing problem (VRP), and Park and Kim [47] documented a range of mathematical approaches in their review of twenty-nine articles on the school bus routing problem (SBRP). All of these articles viewed the SBRP from an operational perspective as a discrete VRP between bus stops and schools. The majority of SBRP research has focused on one of two objectives: minimizing bus travel distance (or the associated cost) or minimizing the number of buses required (as a proxy for the fixed costs of providing the service). However, a few authors have considered other objectives that address maximum route lengths [14, 46], student ride times [4, 38, 59], or student walking distance to a bus stop [5]. All these works treat demand for transportation as occurring at discrete points (bus stops), usually with a specified number of students at each bus stop. We will discuss the relevant mixed load discrete models at the end of this section, after reviewing relevant literature with continuous approximation models.

Analytical continuous approximation models have long been used to gain strategic insight into the impact of changes in transportation policy [41], and there is considerable literature of their use for freight logistics systems and for public transit systems. However, we are not aware of any continuous approximation models for school bus transportation. Continuous approximation models rely on approximations of expected distances for continuously distributed demand, and key foundational references include Beardwood et al. [3], Daganzo [16, 17], Eilon et al. [23], and Larson and Odoni [37].

Langevin et al. [36] and Ho and Wong [32] provide reviews of research that uses continuous approximation modeling for freight logistics systems. More recent freight transportation publications utilizing continuous approximation modeling expand the approach to specific variants of the vehicle routing problem and examine methods for approximating expected distances or defining service regions (e.g., [26, 27, 29, 45, 62]). For a procedure on how to develop implementable routes from continuous approximation models, see del Castillo [22]. A number of recent publications have considered more realistic cost scenarios and service level implications, including Geunes et al. [30], Sankaran and Wood [56], Jabali et al. [33], Tsao and Lu [61] and Davis and Figliozzi [20].

Szplett [58] reviews the literature on continuous approximation models for public transit. More recent public transit research with continuous approximation models considers fixed networks where the routes and stops do not change with demand [10, 18, 25, 44, 63], flexible networks that allow the routes, stops, or both to change with demand [1, 28, 53, 65], and hybrid networks that combine features of these two systems [11, 39, 52].

Although there is a vast literature on the VRP, there is limited research on the SBRP and little research on discrete models for the mixed load SBRP. In the school bus routing survey by Park and Kim [47], only five of the 29 articles addressed the mixed loading variant of the SBRP, and these are briefly described here. Hargroves and Demetsky [31] demonstrated the benefits of a computer assisted approach to design mixed load bus routes for a semi-rural district. Russell and Morrel [54] addressed a mixed load SBRP with a heuristic that transported students first to the nearest school, and then used an inter-school shuttle system to ferry students to their appropriate school. A case study for special education students showed savings of 11 and 16 % on total mileage and travel times, respectively. Chen et al. [12] provided a multi-phase algorithm with a student "cross-dock" (transshipment point) for a rural district in Alabama. Braca et al. [6] adapted the location-based heuristic for the capacitated VRP from Bramel and Simch-Levi [7] in a cluster-first, route-second approach to solve a mixed load SBRP in New York City. Spada et al. [57] compared heuristic algorithms that focused on student ride times for a mixed load SBRP with two real world problem instances and several large artificial data sets.

Campbell et al. [9] augmented the Park and Kim [47] SBRP review with five new articles for the mixed load SBRP. Thangiah et al. [60] presented a multi-step heuristic algorithm for five rural school districts in Pennsylvania, and also provided a valuable discussion of the practical complexities of school bus routing, including government regulations and reimbursement policies. De Souza and Siqueira [21] applied the algorithm of Braca et al. [6] to ten cities in Brazil to explore the savings from using fewer bus stops. Prasetyo et al. [51] described a GIS-based analysis to develop bus routes in Indonesia. Kim et al. [35] considered a bus scheduling problem that links a set of input non-mixed (single school) bus trips to form possibly mixed load routes for a single bus that visits several schools in sequence. Two optimization-based approaches are compared with a heuristic algorithm. Park et al. [48] modeled the mixed load SBRP as a pickup and delivery problem with time windows with the objective of minimizing the total number of buses. Using a heuristic algorithm, computational results were presented in this paper and in the corrigendum [49, 50] for the benchmark data sets introduced in Park and Kim [47]. Of the 10 papers on the mixed load SBRP, the objectives pursued are primarily to minimize the number of buses and minimize the bus travel distance.

3 Strategic Continuous Approximation Models

For our strategic analysis, we use continuous approximation modeling to analyze the benefits of mixed trips for school bus transportation. We first formulate models for the expected travel distance of mixed and non-mixed school bus routes for an idealized school district containing two levels of schools (e.g., lower schools and upper schools). Ellegood et al. [24] derived a more general model for multiple levels of schools. We consider the morning bus trips that pick up students from bus stops and deliver them to schools. The afternoon routing problem of delivering students from the schools to their bus stops can be handled similarly by reversing the directions of travel.

Suppose that each bus completes two trips in the morning, one for each level of school. We assume that the upper level schools are served on the first bus trip, with the lower level schools served on the second trip. The first trip starts at the bus depot, picks up students at bus stops and delivers them to their upper school. The second trip is similar, but it starts from an upper school, rather than the depot. (Extension to more than two schools is straightforward; see [24].) Once the bus has completed its morning trips, it returns to the bus depot to wait for the afternoon trips to bring students home from school.

A non-mixed bus trip picks up students attending only a single school and Fig. 1a shows two non-mixed bus trips where the first trip serves the upper level school (denoted U) and the second trip serves the lower level school (denoted L). Bus stops are shown as diamonds and the two numbers by each bus stop indicate the number of students for the upper and the lower school, respectively. Each bus in this illustration carries 44 students. To transport all students, the non-mixed trips visit six of the bus stops twice since these stops have students for both schools.

A mixed bus trip picks up *all* students at each bus stop, regardless of the student's school. Once the bus is full, it proceeds to deliver the students to each of their assigned schools. Figure 1b shows two mixed bus trips serving the same set of bus stops and students as in Fig. 1a. No stop is visited twice with mixed bus trips and the benefit by having only one trip visit the farthest bus stops is clear. However, non-mixed trips require traveling between schools U and L in Fig. 1b, since both mixed bus trips carry students for both schools. Figure 1a, b shows how the total bus travel distance can be reduced with mixed routing as long as schools are not too far apart. With non-mixed trips in Fig. 1a, the two buses make a total of 13 stops at bus stops and two stops at schools; with mixed trips in Fig. 1b, the two buses visit a total of only 7 bus stops,

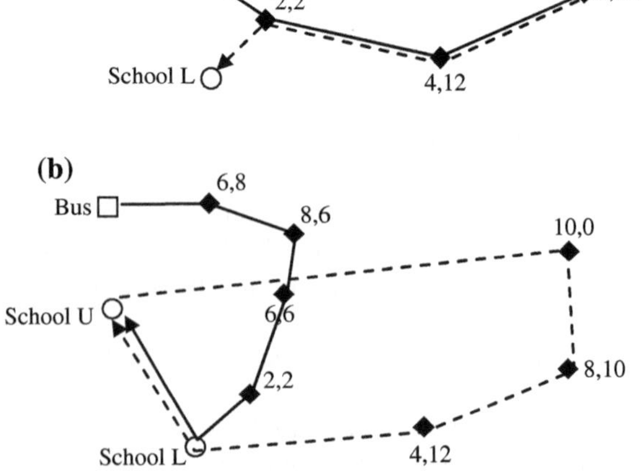

Fig. 1 **a** Two non-mixed bus trips: The *solid trip* is the first trip (for school U) and the *dashed trip* is the second trip (for school L). **b** Two mixed bus trips: The *solid trip* is the first trip, the *dashed trip* is the second trip, and each trip picks up all students at a stop

but make 4 visits to schools. Thus, the main advantages of mixed bus trips is that each bus stop is visited only once (assuming a bus stop does not have more students than can fit on the bus) and fewer stops are required to fill the bus. However, the main disadvantage stems from the travel at the end of the trip between schools with the bus only partially full.

With mixed bus trips, we assume the number of students at a bus stop is less than the bus capacity. (Otherwise, a bus could be filled by visiting one stop, leaving the remaining students to be picked up on another mixed bus trip.) In our model, the mixed bus trips visit the required schools in sequence at the end of the trip, without picking up any students between schools. Models for alternative mixed trip policies (e.g., when not all students at a stop are picked up, or when students are picked up between the schools at the end of the trip) are left for future research.

3.1 Expected Distance for Non-mixed School Bus Trips

Each non-mixed bus trip serves a single school and consists of three components: (i) travel from the trip origin (the bus depot or a school) to the first bus stop; (ii) travel from the first bus stop to the last bus stop while picking up students, and (iii) travel from the last bus stop on the trip to the school. The expected distance for each of these components of travel can be formulated using continuous approximation modeling that treats discrete locations (bus stops or schools) with a continuous spatial density over the service region to derive expected distances. We assume the school district is a compact region A, covering an area A, and use L and U to indicate lower and upper level schools, respectively. Let L be the set of indices for lower level schools, where $L = \{1, 2, \ldots, |L|\}$ and U be the set of indices for upper level schools, where $U = \{|L + 1|, \ldots, |L| + |U|\}$. Then, let set $J = L \cup U$ be the set of indices of all schools. The subscript j denotes a particular school. Each school is assumed to serve a compact subregion of A, where school j serves region A_j of area A_j, and the school subregions for each level fully cover A:

$$\Sigma_{j \in L} A_j = \Sigma_{j \in U} A_j = A. \tag{1}$$

As is common in continuous approximation modeling, we assume the density of bus stops is slowly varying over the region A_j for each school. As in Campbell [8], the expected distance of the first travel component (from the origin to the first bus stop) for a non-mixed bus trip serving school j, is given by

$$D_1^N = K_0(\theta_j)\sqrt{A_j}, \tag{2}$$

where θ_j is the distance from the origin of the bus trip to the centroid of A_j and $K_0(\theta_j)$ is a factor that depends on θ_j and the metric. The K_0 factor depends on the distance between the origin or destination and the region A_j, the shape of A_j, and the distance metric. We use the Euclidean metric throughout our analyses. In the special case where the bus trip origin is in the center of the region being served, then $\theta_j = 0$ and $K_0(\theta_j) \approx 0.383$. When the bus trip origin is not in the center of the region being served (i.e., $\theta_j \neq 0$), then the K_0 factors can be determined when region A_j has a regular shape such as a square, rectangle or circle from equations in Chap. 8 of Eilon et al. [23]. For more general shapes and when the trip origin is outside the region A_j, then Vaughan [64] provides the expression:

$$K_0(\theta_j) = \theta_j \left[\frac{1}{\sqrt{A_j}} + \frac{\sqrt{A_j}}{8\pi\theta_j^2} \right]. \tag{3}$$

The expected travel distance for the third travel component (from the last bus stop to the school) for a non-mixed bus trip serving school j is formulated similarly as

$$D_3^N = K_0(\omega_j)\sqrt{A_j}, \tag{4}$$

where ω_j is the distance from school j to the centroid of A_j.

The expected travel distance for the second travel component (while picking up students between the first and last bus stop) for a non-mixed bus trip serving school j is formulated as the product of the number of stops on the trip, m_j, and a "peddling" factor $K_1(m_j)$ divided by the square root of the density of stops (from [8]),

$$D_2^N = m_j \frac{K_1(m_j)}{\sqrt{\frac{N_j}{A_j}}}, \tag{5}$$

where N_j is the number of bus stops for school j. The peddling factor $K_1(m_j)$ for the Euclidean metric is given in Table 1 (from [8], based on formulae in [16, 17]).

The expected total distance for our case with two morning bus trips can be determined using Eqs. (2), (4) and (5) twice: first for the trip starting at the depot and serving an upper level school, and then for the trip starting at an upper school and serving the lower level school. We assume the bus depot is located at the center of the school district (region A) and the schools of each level are similar and centrally located in compact regions. We use subscript U or L to denote the level of school, so N_U and N_L are the average number of upper and lower level bus stops per school,

Table 1 Trip length peddling factors for multi-stop trips

No. of stops, m	$K_1(m)$
1	0
2	0.73
3	0.68
4	0.63
5	0.60
≥6	0.57

respectively. Also, let E_U and E_L denote the number of upper and lower level schools, respectively.

If there was the same number of students (i.e., bus riders) for each school level, then there would be an equal number of bus trips for each level, However, in general the number of students is not the same for each level, so one level will require more trips than the other, which influences the number of buses required. For example, suppose 12 bus trips are required to transport all students for the upper level schools, but only 8 bus trips are required for the lower level schools. Using 12 buses to complete all the upper level trips simultaneously, would leave 4 unused buses when completing the 8 lower level trips. A similar situation holds if fewer trips are required at the upper level. Because one important goal for school bus transportation is to reduce the total number of buses required, we consider the situation with the minimum number of buses, so each bus makes two trips. Thus, in the example above, only 10 buses would be used, with two buses making two trips for the upper level schools. Adopting such a policy may require adjusting school start times to try to equalize the number of students at each start time, or allowing bus riders to wait longer at their schools before classes begin. More broadly, a large imbalance in the number of students at different levels of schools may lead a school district to consider strategically reorganizing grade levels between buildings and/or adjusting school start times.

In the following analyses we use the minimum number of buses and allow the number of trips for each school level to differ. Let P to denote the total population of bus riders for all schools, which is modeled as a slowly varying spatial density of students (number of students per unit area) for each school. We use a fleet of homogeneous school buses, each with effective capacity C students. If buses travel full for a portion of the trip, then the minimum number of bus trips required is the smallest integer greater than P/C, or P/C if we ignore the fractional component. We let f_U and f_L denote the fraction of these students for the upper and lower level schools, respectively. If the density of bus riders for a particular level of school varies slowly over the district, then the regions served by the different schools of that level will be approximately the same size. Of course, in practice the regions served by different schools will vary based on school sizes, population dynamics, geographic and political boundaries, etc. (and these also change over time). Finally, we assume, as is common, that there are more lower schools than upper schools.

First consider when the population for the upper schools is greater than or equal that for the lower schools, i.e., $f_U \geq 0.5$. There will be $f_U P/C$ upper school trips starting from the center of the school district, so

$$D_1^N \cong 0.383\sqrt{A}.$$

These trips end at an upper level school so

$$D_3^N \cong 0.383\sqrt{A/E_U}.$$

For the travel distance picking up students, note that the average number of stops per trip can be written as the bus capacity divided by the number of upper level students per stop:

$$m_u = \frac{c}{\frac{f_U P}{N_U E_U}} = \frac{C N_U E_U}{f_U P}. \quad (6)$$

From (5) and (6), along with A_U = A/E_U, we have

$$D_2^N = m_U \frac{K_1(m_U)}{\sqrt{\frac{N_U}{A_U}}} = \frac{C}{f_U P}\sqrt{N_U E_U}\sqrt{A}. \quad (7)$$

The total expected distance for the $f_U P/C$ upper school trips to serve E_U schools is then given by the sum of Eqs. (2), (4) and (7)

$$R_U^N \cong \left[\frac{f_U P}{C} 0.383\left(1 + \frac{1}{\sqrt{E_U}}\right) + \sqrt{E_U} K_1(m_U)\sqrt{N_U}\right]\sqrt{A}. \quad (8)$$

There are also $(1-f_U)P/C$ lower school trips whose distance is given in an analogous fashion, but where an upper school is the trip origin. If this origin is in the center of the region it serves of area A/E_U, then the distance for the lower school trips is

$$R_L^N \cong \left[\frac{(1 - f_U)P}{C} 0.383\left(\frac{1}{\sqrt{E_U}} + \frac{1}{\sqrt{E_L}}\right) + \sqrt{E_L} K_1(m_L)\sqrt{N_L}\right]\sqrt{A}. \quad (9)$$

Equation (9) is analogous to Eq. (8) with L replacing U, except for the first component of travel where the upper school, rather than the depot, serves as the origin. To return all buses to the depot adds a distance of

$$0.383\sqrt{A}\frac{P}{2C},$$

so the total distance when $f_U \geq 0.5$ is

$$R_U^N + R_L^N \cong \left[\frac{P}{C} 0.383\left[\frac{1}{2} + \frac{1}{\sqrt{E_U}} + \frac{1}{\sqrt{E_L}} + f_U\left(1 - \frac{1}{\sqrt{E_L}}\right)\right] + \sqrt{E_U} K_1(m_U)\sqrt{N_U} + \sqrt{E_L} K_1(m_L)\sqrt{N_L}\right]\sqrt{A}. \quad (10)$$

Now consider when the population for the upper schools is less than for the lower schools so $f_U < 0.5$. There will be $f_U P/C$ upper school trips whose distance is given by Eq. (8), and these buses will also make $f_U P/C$ lower school trips with a distance similar to (9), but with the population $f_U P/C$, instead of $(1-f_U)P/C$. Because $f_U < 0.5$, there will be an additional $(1–2f_U)P/C$ trips just for the lower schools that will start at the depot and whose distance is given similar to (8). With the return to the depot of all buses, the net result is that the total distance when $f_U < 0.5$ is

$$R_U^N + R_L^N \cong \left[\frac{P}{C}0.383\left[\frac{3}{2} + \frac{1}{\sqrt{E_L}} - f_U\left(1 - \frac{2}{\sqrt{E_U}} + \frac{1}{\sqrt{E_L}}\right)\right] + \sqrt{E_U}K_1(m_U)\sqrt{N_U} + \sqrt{E_L}K_1(m_L)\sqrt{N_L}\right]\sqrt{A}. \quad (11)$$

3.2 Expected Distance for Mixed School Bus Trips

For mixed school bus trips, each bus stop is visited only once and all students at that stop are picked up. (Each bus stop has students associated with only a single school at each level.) Each mixed bus trip will then visit both an upper and a lower level school to deliver students. The number of bus stops with mixed bus trips, denoted Φ^{M}, is likely to be less than the number of bus stops with non-mixed trips, denoted Φ^{N}, where the total number of bus stops with the non-mixed policy is just the sum of the bus stops for each type of school:

$$\Phi^{\mathrm{N}} \cong N_U E_U + N_L E_L, \quad (12)$$

where the approximation is good when the schools at each level have approximately the same number of stops. In practice, the same physical location is often used as a bus stop for different levels of schools, so many of the bus stops coincide, and in general $\max\{N_U E_U, N_L E_L\} \leq \Phi^{\mathrm{M}} \leq \Phi^{\mathrm{N}}$. If buses are full at some point on the route, then the number of mixed trips can be approximated by Φ^{M}/m, where m is the average number of stops on a mixed bus trip, and no subscript is needed since bus stops are no longer distinguished by school or school level.

Each mixed bus trip serves both a lower level school and an upper level school, and includes three components: (i) travel from the trip origin (the bus depot or a school) to the first bus stop; (ii) travel from the first bus stop to the last bus stop while picking up students, and (iii) travel from the last bus stop on the trip to both schools to deliver students. The expected distance of a mixed school bus trip is formulated similar that for non-mixed bus trips, but the second component requires fewer stops since more students are picked up at a stop, and the third component includes additional travel between the upper and lower school at the end of the trip.

The expected travel distance of the first component (from the origin to the first bus stop) of a mixed bus trip is

$$D_1^M = K_0(\theta)\sqrt{A}, \quad (13)$$

where θ is the distance from the origin of the bus trip to the centroid of A. The expected travel distance for the second component of a mixed bus trip is similar to Eq. (5),

$$D_2^M = m\frac{K_1(m)}{\sqrt{\frac{\Phi^M}{A}}}. \tag{14}$$

The expected travel distance for the third component for a mixed bus trip is

$$D_3^M = K_0(\omega)\sqrt{A_\omega} + dist\,UL, \tag{15}$$

where ω is the distance from the first school visited to the centroid of its school region and *distUL* is the distance between the upper and lower school visited at the end of the trip.

The expected total distance for our case with two morning bus trips can be determined using Eqs. (13)–(15): first for the trip starting at the depot, and then for the trip starting at a school. To illustrate the mixed trip distance model with two levels of schools (upper and lower schools), we assume the lower level school is visited first. Therefore, the first trip ends at the upper school, where the second trip begins (instead of at the depot). The expected distance for a mixed trip is formulated similar to that for a non-mixed trip as above, but the fraction of upper and lower school students is no longer a factor as all students at a stop are picked up with the single bus visit. Thus, the expected distance of a mixed bus trip starting at the depot is

$$RM_1 = \left[0.383\left(1 + \frac{1}{\sqrt{E_U}} + \frac{1}{\sqrt{E_L}}\right) + mK_1(m)\frac{1}{\sqrt{\Phi^M}}\right]\sqrt{A}. \tag{16}$$

The second mixed trip starts at an upper school, so when this is in the center of a region of area A/E_U, then the distance for this second mixed trip is

$$RM_2 = \left[0.383\left(\frac{2}{\sqrt{E_U}} + \frac{1}{\sqrt{E_L}}\right) + mK_1(m)\frac{1}{\sqrt{\Phi^M}}\right]\sqrt{A}. \tag{17}$$

When each bus makes two trips (one trip for each school level), then the minimum number of buses required would be approximately $P/2C$, and the total travel distance for all trips, with half from (16) and half from (17), is

$$R^M = \left[0.383\frac{P}{C}\left(0.5 + \frac{1.5}{\sqrt{E_U}} + \frac{1}{\sqrt{E_L}}\right) + K_1(m)\sqrt{\Phi^M}\right]\sqrt{A}. \tag{18}$$

To return the buses to the depot would add an average distance of

$$0.383\sqrt{A} \times \frac{P}{2C}$$

and make the total distance for all mixed bus trips:

$$R^M = \left[0.383\frac{P}{C}\left(1 + \frac{1.5}{\sqrt{E_U}} + \frac{1}{\sqrt{E_L}}\right) + K_1(m)\sqrt{\Phi^M}\right]\sqrt{A}. \tag{19}$$

3.3 Comparison of Mixed and Non-mixed Routing

The analytical models of non-mixed and mixed school bus travel distance derived above depend on only a few basic parameters that describe the setting and the operations of the school district. For non-mixed trips, the expected distance also depends on the distribution of students between the two school levels (i.e., f_U), while for mixed trips the distribution of students between different school levels is not a factor as all students at a stop are picked up together. However, with mixed trips the degree to which stops are shared between schools and the distance between the upper and lower schools are important factors (see Eq. (19)). To help identify the conditions under which mixed trips are likely to be beneficial, we analyzed how the savings from mixed trips are affected by key problem parameters.

For our general analysis, consider a baseline school district serving a geographic area of $A = 30$ square miles with $E_U = 4$ upper level schools and $E_L = 8$ lower level schools, where each school is centrally located in a compact region. There are $P =$ 2,000 student bus riders with $N_U = 100$ school bus stops per upper level school and $N_L = 50$ school bus stops per lower level school. Buses have capacity $C = 50$, so (at least) 40 bus trips are required to transport all 2000 students.

The total distance using non-mixed trips is given by Eq. (10) or (11), depending on whether the majority of students are for the upper or lower schools. The total distance using mixed trips is given in Eq. (19). To assess how the estimated savings from mixed trips are affected by proportion of shared bus stops, we analyzed varying degrees of sharing of stops with a different mix of students for the two school levels. In our baseline school district there are $\Phi^{\mathrm{N}} = 800$ non-mixed bus stops with 400 for each level of school. With mixed trips, when 100 % of upper and lower school stops are shared, then there are $\Phi^{\mathrm{M}} = 400$ mixed bus stops. Similarly, when 50 % of upper and lower schools stops are shared, then there are $\Phi^{\mathrm{M}} \approx 533$ mixed bus stops; and when 0 % of upper and lower school stops are shared, then there are $\Phi^{\mathrm{M}} = 800$ mixed bus stops.

Figure 2 plots the expected total travel distance of the mixed and non-mixed trips for the baseline school district, but with varying numbers of bus stops. The dashed curve shows the expected distance for non-mixed trips, while the other three curves show the expected distance for mixed trips with 0, 50 and 100 % shared bus stops. The three mixed trip curves document the reduced travel distance with an increasing degree of sharing of the bus stops. This figure also illustrates how decreasing the number of bus stops decreases the travel distance, especially with non-mixed trips where the slope of the distance curve is steeper than for the distance curves for mixed trips. The intersection of the dashed curve for non-mixed trips with the other curves indicates the transition from non-mixed to mixed trips. With few bus stops, non-mixed trips (dashed line) provide the lowest travel distance. However, as the number of bus stops increases, then mixed trips are increasingly preferred, especially when there is a large percentage of shared stops. For the baseline case of 800 bus stops, mixed trips are better even with 0 % shared stops. Note that consolidation of bus stops to produce fewer stops (which reduces travel distances), usually produces more shared stops, which further reduces travel distances for mixed routes.

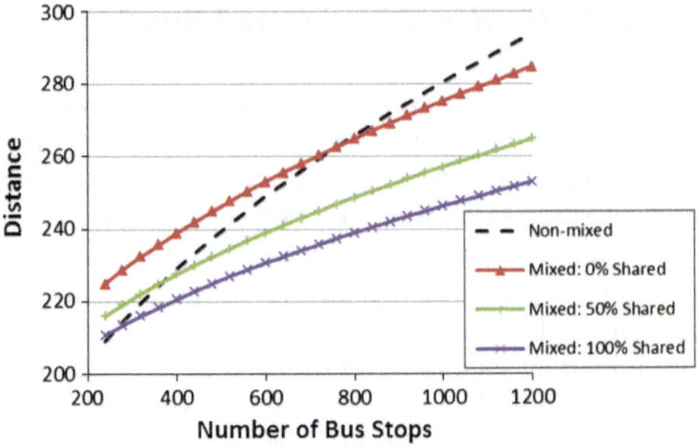

Fig. 2 Travel distance as a function of the number of bus stops

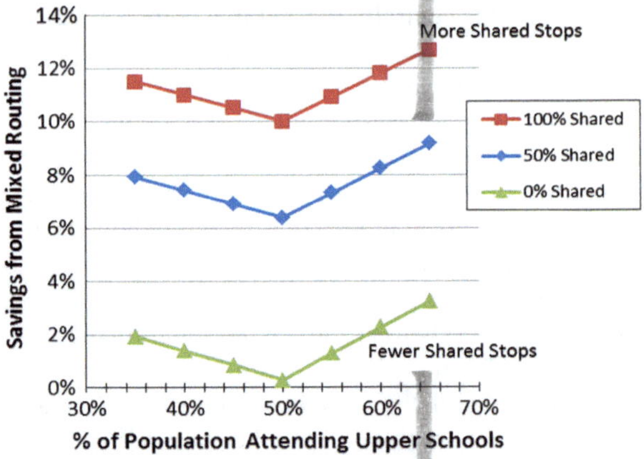

Fig. 3 The impact of shared stops between school levels

Figure 3 shows the percentage savings in travel distance from mixed trips as a function of the fraction of students attending upper schools. The three curves in the figure show the savings with 0, 50 and 100 % shared stops. For the baseline school district with 100 % shared stops, mixed bus trips reduce travel distance by about 10–13 %, compared to a non-mixed trips, depending on the distribution of the population between the school levels. The benefits from a mixed policy decrease as the degree of sharing of bus stops declines, but increase slightly with the deviation from an even mix of students between upper and lower schools. Note that with a 50–50 mix of students and 0 % shared stops, there is essentially no benefit from mixed trips.

Figure 4 shows the percentage savings in travel distance from mixed trips as a function of the fraction of students attending upper schools for different numbers of students in the district, while keeping the number of bus stops constant and with 50 % shared stops. These results are also with the baseline of four upper level schools and eight lower level schools. The curves in Fig. 4 show the how the savings increase

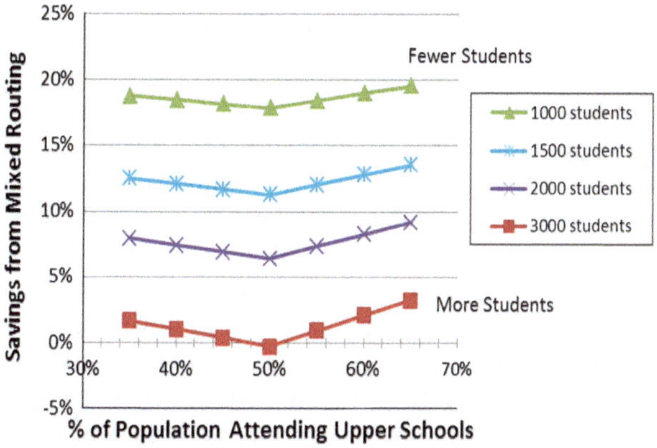

Fig. 4 The impact of the number of students within a district: 50 % shared stops

up to almost 20 % with only 1000 students. This is due to the increasing effectiveness of the mixed trips in visiting fewer stops to fill the bus; non-mixed trips must visit many more stops to fill the bus when each stop has very few students. The results are very similar with 100 % shared stops, but savings are 3–5 % greater than in Fig. 4.

The results as exemplified above suggest how mixed trips can reduce travel distance for a school district. In addition to the influences highlighted above, the magnitude of the benefits from mixed routing depends on the travel distance between the schools at the end of each trip. The continuous approximation models use expected travel distances based on the schools being randomly located in the district and centrally located relative to the students they serve. Therefore, when there are only a few schools, the expected distances between schools can be quite large. In practice, several schools are often located close together to provide efficiencies in operations and joint access to facilities (such as athletic fields). In these cases, the travel distance between schools could be quite small and the benefits from mixed trips would be even greater than suggested above. See Ellegood et al. [24] for a more detailed analysis of school locations.

The results of the strategic modeling suggest that mixed school bus trips have the potential to reduce travel distances, especially in larger districts with few students per stop and when a large percentage of stops are shared. These are characteristics of rural school districts where students live far apart; hence there are few students per bus stop and the bus stop density is low, and districts cover large geographic regions. While the continuous approximation modeling is useful to suggest that mixed trips may be beneficial, this approach does not provide implementable routes. So to complement the strategic analysis, we develop a heuristic to determine mixed load bus routes where bus stops and schools are specified as discrete points.

4 Discrete Mixed Trip Bus Routing Algorithm

A discrete school bus routing problem consists of a set of students at bus stops, a set of schools, a depot (sometimes co-located with a school), a set of operational parameters, and a set of capacitated buses. Diverging from the continuous approximation approach, the discrete problem models the depot, the bus stops, and the schools as discrete points in a service area. The distances between these points may be calculated using the point coordinates and a particular metric (e.g., straight line distances), via shortest paths on a network, or from travel on the underlying road network. Travel times between these points are then estimated based on the travel distance. The goal of our discrete model for mixed trip school bus routing is to create a near-optimal set of bus routes that ensure that each student is delivered to their school within a specified time window. We also include a riding time limit for each student to prevent the creation of excessively long routes. Thus, unlike the continuous approximation modeling, in this section the time windows for delivery to the schools will be respected, so the solution involves both routing and scheduling of the bus trips. The most common criteria pursued in school bus routing problems are minimization of total travel distance or the total number of buses required. Our hybrid algorithm addresses both of these criteria, though with greater attention to minimizing the total travel distance.

We have developed a three-phase heuristic solution algorithm for the mixed trip school bus routing problem in the spirit of Thangiah et al. [60], where individual heuristic components for route creation and route improvement are used in sequence. The first phase of the heuristic is a mixed load implementation of the Savings Method [13], followed by two distance-based improvement heuristics. The second phase of the heuristic seeks to reduce the number of bus trips by combining shorter bus trips when favorable, and it also incorporates the distance improvement procedures utilized in phase one. This phase is repeated until no improvement occurs. The output of this phase is a set of possibly mixed load bus trips that are then linked together in the third phase of the heuristic to reduce the number of buses required. In each phase of the algorithm, no improvement is accepted unless the bus capacity and maximum student ride times are respected. The algorithm is summarized in Fig. 5.

The construction heuristic of phase 1 begins with a modified implementation of the Clarke and Wright savings heuristic, similar to the approach in Russell and Morrel (1986). In this heuristic, a positive "savings" value is generated for each pair of bus stops and these are sorted in decreasing order. Preliminary bus trips are created which include the trip origin, a bus stop, and the destination school for the students at that stop. These preliminary trips are iteratively merged in decreasing order of savings until either all of the bus stops have been assigned to a trip, or all savings values have been considered. The mergers allow one trip to be added to the end of the other, and whenever there are two schools involved these schools are visited in sequence at the end of the trip. Thus, with two different schools to be visited there are four possible configurations of the resulting merged bus trip depending on which original trip comes first and which school is visited first. Figure 6 illustrates this procedure

Phase 1: Construct a set of mixed load bus trips

__1a.__ Modified Clark & Wright trip construction heuristic

__1b.__ Two-Opt distance improvement heuristic

__1c.__ Stop Exchange distance improvement heuristic

Phase 2: Reduce the number of trips and total distance

While the set of trips is still changing: loop

Trip Remove bus reduction heuristic

Two-Opt distance improvement heuristic

Stop Exchange distance improvement heuristic

Two-to-one bus reduction heuristic

Two-Opt distance improvement heuristic

Stop Exchange distance improvement heuristic

Update the set of bus trips

End

Phase 3: Link bus trips to form bus routes

Fig. 5 Three-phase mixed load SBRP heuristic

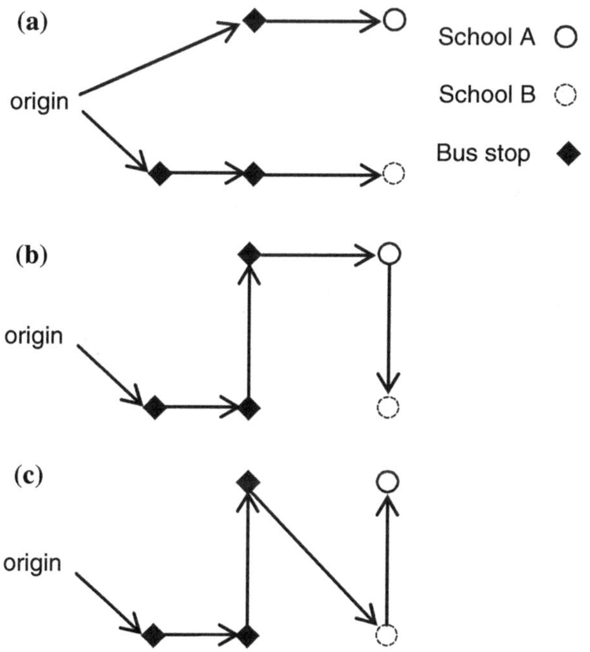

Fig. 6 Savings procedure for mixed trips. **a** Original two trips. **b** One trip: School A before School B. **c** One trip: School B before School A

with two of the four possible configurations for merging trips that have two different destination schools.

Figure 6a shows two original trips from the same origin, but to different schools. Figure 6b, c shows two possible merged trips where the (lower) two-stop original trip precedes the (upper) one-stop original trip. In Fig. 6b, the two trips are merged to form a single trip visiting school A prior to school B. Figure 6c has the same sequence of stops from merging the trips, but here school B is visited before school A. The other two configurations possible for savings calculations with mixed trips insert the single bus stop of the upper original trip before the two stops in the lower original trip.

Once the savings procedure creates initial trips, they are processed with two improvement heuristics. The well-known two-opt heuristic [15] is an improvement procedure that removes two non-adjacent arcs in a trip and replaces them with two new arcs that recreate the trip. The stop exchange heuristic attempts to improve the preliminary set of bus trips by switching two bus stops between a pair of trips. The exchange is accepted if the resulting total distance for the two new trips is less than the total distance for the two initial trips and if both new trips are feasible. A bus stop can be moved to a new trip only if that trip is already visiting the school associated with the stop being switched. Note that the heuristic does not distinguish between mixed and non-mixed trips when performing exchanges, but does ensure the feasibility requirements are respected. Both of these improvement heuristics are used to improve the initial trips. See Campbell et al. [9] for details on the implementation.

The second phase of the heuristic seeks to reduce the number of trips and the total travel distance by iterating through a series of trip reduction and distance reduction heuristics. First, a trip reduction heuristic is employed with the goal of eliminating any trip that utilizes less than half of the capacity of a bus by relocating their stops to other trips. Stops are removed from short trips and relocated to other trips that have the available capacity and will not violate the trip time limit or the school time window. If an entire trip can be eliminated, then the number of buses required is reduced. This is followed by application of the two-opt heuristic and the stop exchange heuristic, as described above.

Next, the two-to-one bus reduction heuristic is used to merge two underutilized trips into a single (longer) trip. This heuristic reorders the stops from two trips to create a single trip visiting the same school(s). As before, the bus capacity, trip travel time limit and school time windows must be obeyed in the resulting trip. This procedure is followed by application of the two-opt heuristic and the stop exchange heuristic, as described above. At the end of the second phase of the heuristic, the algorithm has created a set of bus trips where each originates at the depot, visits one or more bus stops, and then visits the necessary schools.

The third and final phase of the heuristic is designed to sequentially link bus trips to form bus routes in an effort to reduce the number of buses required. Thus, once a trip from phase two has been completed and a bus is empty at a school, rather than having the bus return to the depot it attempts to travel directly to a bus stop to begin another trip. To accomplish this, we implement a binary integer programming model to minimize the number of buses.

We formulate this trip linking problem on a directed graph $G = (V, A)$ where the vertex set $V = \{\Theta\} \bigcup T$ consists of the depot Θ and the set of vertices T that correspond to the bus trips. Each trip is represented as a single vertex, and the distance d_{ij} represents the distance from the last stop of trip $i \in T$, which is at a school, to the first bus stop of trip $j \in T$. The asymmetry of the distance matrix ($d_{ij} \neq d_{ji}$) is an important aspect of this problem, because the first and last stops of two trips i and j are generally not the same. To determine the trip linkages, a binary decision variable $X_{ij} \in \{0, 1\}$ equals one if bus trip $i \in T$ immediately precedes bus trip $j \in T$ on a bus route, and 0 otherwise. Note that when

$$\Sigma_{j \in T} X_{ij} = \Sigma_{j \in T} X_{ji} = 0,$$

then trip i is not linked to any other trip.

Trip Linking Model

$$maximize\ \Sigma_{i \in T} \Sigma_{j \in T} X_{ij} \tag{20}$$

subject to

$$\Sigma_{i \in T} X_{ij} \leq 1 \qquad \forall j \in T \tag{21}$$

$$\Sigma_{j \in T} X_{ij} \leq 1 \qquad \forall i \in T \tag{22}$$

$$X_{ii} = 0 \qquad \forall i \in T \tag{23}$$

$$X_{ij} \in \{0, 1\} \qquad \forall i, j \in N \tag{24}$$

Objective (20) maximizes the total number of trip linkages in the solution. This minimizes the number of buses required (for the set of input trips) because each linkage reduces the number of trips by one. Constraint (21) ensures that each trip can have at most one immediate predecessor trip and constraint (22) ensures that each trip can have at most one immediate successor trip. Constraint (23) prevents trips from being linked to themselves, and constraint (24) establishes a binary restriction on the decision variable. The trip linking model takes as input a set of mixed bus trips created in the first two phases of the heuristic. The output is a set of mixed load routes to be driven by a bus, where each route consists of one or more mixed load trips.

To test the three-phase heuristic algorithm, we considered the benchmark mixed load school bus routing problems in Park et al. [48]. We solved 16 instances from their "random" data set where the number of schools ranges from 6 to 100 and the number of bus stops ranges from 250 to 2000. The depot is centrally located, while the schools and bus stops are located randomly across a 20 × 20 mile square region. Each school has a randomly chosen earliest start time between 7:00 and 11:00 a.m., and a time window for arrival of buses between 10 and 30 min long. The number of students at each bus stop was generated randomly and the capacity of the bus is set at 66. The loading and unloading (service) times for students are given by the regression models in Braca et al. [6]. Note that the random assignment of school start times is rather unrealistic as a school district will typically coordinate school start times to facilitate bus transportation.

Table 2 Results on random benchmark data sets with 45 min maximum ride time

			Park et al. [48–50]		Three phase heuristic	
Data set	Stops	Schools	Distance	#Buses	Distance	#Buses
RSRB01	250	6	11,954,142	**30**	**8,351,909**	34
RSRB02	250	12	12,262,879	29	**9,427,990**	**26**
RSRB03	500	12	21,834,065	56	**17,590,183**	**54**
RSRB04	500	25	24,172,922	59	**14,918,581**	**45**
RSRB05	1,000	25	40,008,910	**98**	**31,228,776**	108
RSRB06	1,000	50	43,981,860	89	**28,956,790**	**77**
RSRB07	2,000	50	77,579,541	**154**	**56,239,960**	174
RSRB08	2,000	100	82,475,521	**157**	**66,344,133**	165
Average	938	35	39,283,730	**84**	**29,132,290**	85

Table 3 Percent savings on random data sets with 45 min maximum ride time

	Percent savings		
Data set	Distance (%)	#Buses (%)	CPU time (s)
RSRB01	30	−13	31.3
RSRB02	23	10	16.2
RSRB03	19	4	78.7
RSRB04	38	24	59.7
RSRB05	22	−10	398.3
RSRB06	34	13	148.7
RSRB07	28	−13	1868.9
RSRB08	20	−5	2678.6
Average	27	1	660.1

Tables 2, 3, 4 and 5 display the results for the algorithm with a maximum riding time of 45 and 90 min for any student. The first column of the tables displays the name of the data set. The second and third columns of Tables 2 and 4 indicate the number of bus stops and schools. The fourth and fifth columns of Tables 2 and 4 provide the travel distance and the number of buses from Park et al. [49, 50], the best known results for this dataset in the literature. Park et al. [48] presented a heuristic to minimize the number of buses required and we obtained their detailed results for each bus route, and then generated the corresponding travel distance shown in column 4. Columns 6 and 7 of Tables 2 and 4 provide the travel distance and the number of buses from our heuristic. Tables 3 and 5 show the percentage savings in distance and buses from the three-phase heuristic, relative to the values from Park et al. [48–50]. The last column of Tables 3 and 5 provides the cpu time for the three-phase heuristic, which seems quite reasonable given that routes are designed at the beginning of the school year. Bold numbers in the table designate the minimum values.

Figure 7 plots the percentage savings in travel distance and buses for the three-phase heuristic for the problems in Tables 2, 3, 4 and 5. These results show how the heuristic consistently finds bus trips with shorter total travel distance, averaging 27 and 19 % shorter for the problems with a 45 and 90 min ride time, respectively.

Table 4 Results on random benchmark data sets with 90 min maximum ride time

			Park et al. [48–50]		Three phase heuristic	
Data set	Stops	Schools	Distance	#Buses	Distance	#Buses
RSRB01	250	6	10,707,222	**27**	**8,540,572**	33
RSRB02	250	12	12,079,821	**23**	**9,427,990**	26
RSRB03	500	12	19,250,780	**47**	**17,590,183**	54
RSRB04	500	25	20,846,180	46	**14,918,581**	**45**
RSRB05	1,000	25	35,792,599	**79**	**31,228,776**	108
RSRB06	1,000	50	38,847,232	**72**	**28,956,790**	77
RSRB07	2,000	50	69,245,736	**134**	**56,239,960**	174
RSRB08	2,000	100	78,390,582	**142**	**66,344,133**	165
Average	938	35	35,645,019	**71**	**29,155,873**	85

Table 5 Percent savings on random data sets with 90 min maximum ride time

	Percent savings		
Data set	Distance (%)	#Buses (%)	CPU time (s)
RSRB01	20	−22	48.9
RSRB02	22	−13	16.2
RSRB03	9	−15	78.7
RSRB04	28	2	59.7
RSRB05	13	−37	398.3
RSRB06	25	−7	148.7
RSRB07	19	−30	1868.9
RSRB08	15	−16	2678.6
Average	19	−17	662.3

In terms of the number of buses required, the three-phase heuristic in aggregate saves 1 bus (<1 %) with the 45 min ride time limit, but requires 17 % more buses with the 90 min time limit. The strong performance for travel distance relative to Park et al. [49, 50] is not surprising given our primary focus on minimizing travel distance with mixed trips, compared to their focus on minimizing the number of buses. However, the attention in the three-phase heuristic to minimizing the number of buses does pay dividends, especially with the tighter ride time limits. Figure 7 does suggest the tradeoff between buses and travel distance as greater savings in travel distance does seem associated with using more buses. However, care is needed when interpreting these results as they are relative to the solutions in Park et al. [49, 50], whose performance may depend on the problem parameters (number of schools, ride time limit, etc.). These results do show that the three-phase heuristic performs quite well in its primary objective of minimizing travel distance with mixed trips, and also does well to minimize the number of buses required with short ride time limits. However, with the longer ride time limits, the procedure in Park et al. [48] is better able to exploit the additional trip times to reduce the number of buses, though at the expense of greater travel distances.

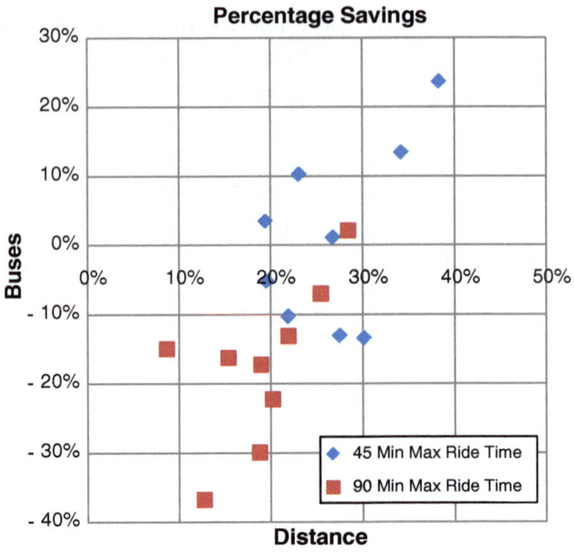

Fig. 7 Percentage savings compared to benchmark results

5 Case Study

This section describes a case study for the Windsor School District (WSD), a semi-rural school district located approximately 25 miles south of St. Louis, Missouri, to demonstrate the applicability of the models and some of the real-world complications in school bus routing. The school district has approximately 3,000 students, 2,300 of whom are transported to and from school by school buses. WSD has five schools; Windsor High School (HS) for grades 9–12, Windsor Middle School (MS) for grades 6–8, Windsor Intermediate School (IS) for grades 3–5, Windsor Elementary School (WE) for grades Pre K-2, and Freer Elementary School (FE) for grades Pre K-3. Note that this includes four levels of schools and an overlap in grade 3 between IS and FE. The school district is approximately a rectangle of 17.1 square miles, measuring 5.25 miles north-to-south and 3.25 miles east-to-west. For this case study, the buses and the student population being served by special-needs buses are excluded. One complication for WSD concerns the students in grade 3. Elementary school students in the southwest quadrant of the district attend nearby school FE for grades Pre K-3, but all other elementary students attend school WE for grades Pre K-2, and then move to the intermediate school IS for grade 3. Another interesting feature of WSD is that schools HS, MS and IS are very close together in the east-central part of the district, with WE and the bus depot also located nearby. Table 6 provides basic data on the grades and bus riders for each school. The mix of schools and overlapping assignments of grades reflects the complexity of real-world school planning.

Currently WSD utilizes mixed routing with 22 buses, with each bus completing two trips in the morning and two trips in the afternoon. We concentrate on the morning trips in the case study. Currently, the first trip for all 22 buses is a mixed trip

Table 6 Schools, grades and bus riders for the Windsor School District

	HS	MS	IS	WE	FE	Total
Grades	9–12	6–8	3–5	Pre-K-2	Pre-K-3	-
Bus rider population	484	806	405	387	219	2,301

for students attending MS and HS, with each trip starting at the bus depot, making an average 8.4 stops picking up approximately 7 students per stop, and then transporting the students to MS first, and then to HS. All second trips for students attending IS or an elementary school begin at HS, with eight of the mixed trips serving the students for IS and FE (with an average of 6.4 students per stop) and the remaining fourteen mixed trips serving students for IS and WE (with an average of 5.2 students per stop).

The current policy in WSD can be analyzed using the continuous approximation travel distance equations developed in this research. To calculate the total expected distance for the current WSD mixed bus trips, we use the actual locations of the bus depot and the schools, but assume that the bus stops are randomly located over the district. The current mixed routing policy for WSD has a total expected travel distance from the continuous approximation models for both sets of 22 trips of 296.6 miles (see [9] for details). In addition to analyzing the current policy for WSD, we also analyzed serving all students with non-mixed trips, where the first set of 22 trips consists of 8 non-mixed HS trips and 14 non-mixed MS trips, and then the second set of 22 trips consists of 10 non-mixed trips for IS, 8 non-mixed trips for WE, and 4 non-mixed trips for FE. This produced a total expected travel distance only 1 % greater than with mixed trips. A closer examination of the travel distance for each school revealed a long distance travel between the schools in the *mixed* trips serving far apart schools IS and FE, due to the start time for IS preceding that for FE. Thus, FE students in the southwest were being transported across the district to IS to meet its earlier start time, before returning back to FE (with the later start time). To eliminate this long travel distance for the FE students, we considered a hybrid strategy where the first 22 trips are mixed trips for HS and MS, but the second trips are split into 14 mixed trips for nearby schools IS and WE, and then 4 non-mixed trips for FE and 4 non-mixed trips for IS. This hybrid strategy produced savings in travel distance of 8.5 %, which shows the benefits of tailoring school bus transportation to the specific details in the school district, especially school locations. These results suggest that replacing the current transportation policy with a non-mixed policy throughout the district will result in a negligible change in the total distance travelled, but using a hybrid policy could result in a noticeable reduction in the total distance travelled (about 8.5 %).

We also applied the three-phase heuristic to develop discrete mixed trip bus routes for WSD. We first geocoded the bus stop and school locations and calculated the road travel distance using the shortest path through the road network between pairs of locations. Then, we analyzed the current bus routes in WSD by modeling each of the routes for the 22 buses to determine the total travel distance was 434.9 miles.

As expected, this is considerably larger than the distance from the continuous approximation models as that used straight-line distances and the heuristic uses shortest paths on the actual road network. Lengthy travel was observed again for back-and-forth travel between FE and IS due to the start time conflict for these schools. We then used the three-phase heuristic to determine the bus routes, and it produced much shorter routes totaling 329.2 miles with only 17 buses. Interestingly, the heuristic produced non-mixed trips routes serving FE and mixed trips serving the other schools, which reflects the hybrid strategy that was suggested to be beneficial by the continuous approximation modeling. These hybrid routes from the discrete heuristic used fewer buses than the current routes in WSD (17 vs. 22), and they respected the 45 min ride time restriction and school start times employed by WSD. However, the routes from the heuristic were of longer duration and carried more students than the routes currently used by WSD, as the heuristic tended to better fill the buses. We acknowledge though that there may be good practical reasons to design somewhat shorter routes, so that some unexpected delays en route (e.g., traffic congestion, longer than expected loading) can be accommodated without making the students arrive late at school. We also used the three-phase heuristic with an additional restriction that all trips should be non-mixed, and this produced routes requiring 21 buses and a total distance of 354.8 miles, an increase of 7.8 % compared to the mixed trips. So in summary, the results for the discrete model with the three-phase heuristic for the WSD case study are in general agreement with the findings from the continuous approximation modeling in showing the benefits of mixed trips and the value of a hybrid strategy. However, we do underline that caution is needed when comparing the travel distances from the continuous approximation model and the heuristic due to the different ways of measuring distance. Note also that the shortest path distances used in the heuristic may understate actual bus travel distances, because the large buses may not always follow shortest paths on the roads for safety reasons.

6 Conclusions and Future Research

School bus routing is an important part of student transportation and mixed loading is one simple technique that may help improve bus routes and reduce costs for school districts. In this paper, we describe our research using analytical continuous approximation models and a discrete three-phase heuristic to evaluate mixed bus trips. The continuous approximation approach uses very limited data to approximate expected school bus travel distances. These strategic models help to identify conditions under which mixed school bus trips may be beneficial. Results showed that mixed trips are more beneficial when students are sparsely distributed, when there are many bus stops, and when a large percentage of stops are shared. The results also show that mixed routing is more beneficial when the distribution of students between schools is uneven.

With the discrete mixed load bus routing heuristic we generated actual bus routes and compared their performance to best known results on large benchmark problems.

Our approach created routes with considerably less travel distance than another method from the literature, though the number of buses required was often larger. Results showed the tradeoff between minimizing the number of buses, which may require some longer routes than desired, and minimizing travel distance, but with more buses. A case study was presented to demonstrate the applicability of both modeling approaches in practice, and results for the case study were quite similar for both modeling approaches.

A synthesis of results suggests that mixed school bus trips may often be able to reduce travel distances for schools that are not too far apart, and that mixed trips are likely to be more beneficial in rural school districts due to the low density of stops and the fewer students per stop. Further, a hybrid routing strategy may often be desirable, where nearby schools can be served on mixed trips but widely separated schools are served using non-mixed trips. More generally, we have shown how the continuous approximation models and discrete routing algorithms can be used together to provide valuable insights for school bus routing.

We must note that our research focused primarily on modeling the school bus travel distance, as that is an important financial concern. However, actual school bus routes are subject to a variety of complicated practical and local issues, including safety, road type, bus route length, stop location, student age conflicts, walking distance, etc. [43]. Also, many educational issues are relevant for school bus routing, because school bus transportation policies (route lengths, start times, composition of ridership, etc.) may influence educational achievement. This is certainly an issue of importance for school transportation personnel, educators, students and their parents, but it is beyond the scope of this research. So, we must emphasize that while this research has demonstrated the promise from mixed load school bus trips in terms of reducing bus travel distance, these improvements are not guaranteed in any particular setting and extrapolation of results from one district to another is not recommended. However, the continuous approximation modeling approach provides a relatively easy way to assess the potential improvements from mixed trips.

Several promising areas of future research stem from the ideas in this paper. One extension would be to consider different versions of mixed bus trips where students are picked up between the schools at the end of the trip. Another area of future research involves analyzing bus stop consolidation policies, since combining bus stops can reduce the bus travel distance, though at the added expense of students traveling (e.g., walking) farther to reach the bus stop. Fewer bus stops may increase the inconvenience of bus riding and lead to more students choosing alternative means of getting to school (riding a bike, driving or riding with parents or others). Thus, the model choice decision for the trip to and from school is an interesting area where utility models with mode choice may be useful.

Acknowledgments The work was supported by the University of Missouri Research Board.

References

1. Aldaihani MM, Quadrifoglio L, Dessouky MM, Hall R (2004) Network design for a grid hybrid transit service. Transp Res A-Policy 38:511–530
2. American School Bus Council (2013) Environmental benefits. Available from http://www.americanschoolbuscouncil.org/issues/environmental-benefits. Accessed 22 Nov 2013
3. Beardwood J, Halton JH, Hammersley JM (1959) The shortest path through many points. Proc Camb Philos Soc 55:299–327
4. Bennett BT, Gazis DC (1972) School bus routing by computer. Transp Res 6:317–325
5. Bowerman R, Hall B, Calamai P (1995) A multi-objective optimization approach to Urban school bus routing: formulation and solution method. Transp Res A-Policy 29(2):107–123
6. Braca J, Bramel J, Posner B, Simchi-Levi D (1997) A computerized approach to the New York City school bus routing problem. IIE Trans 29:693–702
7. Bramel J, Simchi-Levi D (1995) A location based heuristic for general routing problems. Oper Res 43:649–660
8. Campbell JF (1993) One-to-many distribution with transshipments: an analytic model. Transp Sci 27(4):330–340
9. Campbell JF, Ellegood WA, North JW (2013) Optimizing school bus routing in Missouri. Final report for the University of Missouri Research Board, 7 August 2013
10. Chang SK, Schonfeld P (1991) Multiple period optimization of bus transit systems. Transp Res B-Methodol 25(6):453–478
11. Chang SK, Schonfeld P (1991) Optimization models for comparing conventional and subscription bus feeder services. Transp Sci 25(4):281–298
12. Chen D, Kallsen H, Chen H, Tseng V (1990) A bus routing system for rural school districts. Comput Ind Eng 19:322–325
13. Clarke G, Wright J (1964) Scheduling of vehicles from a central depot to a number of delivery points. Oper Res 12:568–581
14. Corberan A, Fernandez E, Laguna M, Marti R (2002) Heuristic solutions to the problem of routing school buses with multiple objectives. J Oper Res Soc 53:427–435
15. Croes GA (1958) A method for solving traveling salesman problems. Oper Res 6:791–812
16. Daganzo CF (1984) The length of tours in zones of different shapes. Transp Res B-Methodol 18(2):135–145
17. Daganzo CF (1984) The distance traveled to visit N points with a maximum of C stops per vehicle: an analytic model and an application. Transp Sci 18(4):331–350
18. Daganzo CF (2010) Structure of competitive transit network. Transp Res B-Methodol 44:434–446
19. Daganzo CF, Gayah VV, Gonzales EJ (2012) The potential of parsimonious models for understanding large scale transportation systems and answering big picture questions. Eur J Trans Log 1:1–19
20. Davis BA, Figliozzi MA (2013) A methodology to evaluate the competitiveness of electric delivery trucks. Transp Res E-Log 49:8–23
21. De Souza L, Siqueira P (2010) Heuristic methods applied to the optimization school bus transportation routes: a real case. In: IEA/AIE'10 proceedings of the 23rd international conference on industrial engineering and other applications of applied intelligent systems—volume part II, pp 247–256
22. del Castillo JM (1999) A heuristic for the traveling salesman problem based on a continuous approximation. Transp Res B-Methodol 33:123–152
23. Eilon S, Watson-Gandy CD, Christofides N (1971) Expected distances in distribution problems. In: Distribution management: mathematical modelling and practical analysis. Griffin, London
24. Ellegood WA, Campbell JF, North JW (2013) Continuous approximation models for mixed load school bus routing. Working paper
25. Estrada M, Roca-Riu M, Badia H, Robuste F, Daganzo CF (2011) Design and implementation of efficient transit networks: procedure, case study and validity test. Proced Soc Behav Sci 17:113–135

26. Figliozzi MA (2007) Analysis of the efficiency of urban commercial vehicle tours: data collection, methodology, and policy implications. Transp Res B-Methodol 41:1014–1032
27. Francis P, Smilowitz K (2006) Modeling techniques for periodic vehicle routing problems. Transp Res B-Methodol 40:872–884
28. Fu L (2002) Planning and design of flex-route transit services. Transp Res Rec 1791:59–66
29. Galvao LC, Novaes AG, de Cursi JE, Souza JC (2006) A multiplicatively-weighted Voronoi diagram approach to logistics districting. Comput Oper Res 33:93–114
30. Geunes J, Shen Z-JM, Emir A (2007) Planning and approximation models for delivery route based services with price-sensitive demands. Eur J Oper Res 183:460–471
31. Hargroves B, Demetsky M (1981) A computer assisted school bus routing strategy: a case study. Socio-Econ Plann Sci 15:341–345
32. Ho HW, Wong SC (2006) Two-dimensional continuum modeling approach to transportation problems. J Transp Syst Eng Inf Technol 6(6):53–72
33. Jabali O, Gendreau M, Laporte G (2012) A continuous approximation model for the fleet composition problem. Transp Res B-Methodol 46:1591–1606
34. Johnson N, Oliff P, Williams E (2011) An update on state budget cuts. http://www.cbpp.org/cms/index.cfm?fa=view&id=1214. Accessed 9 Feb 2011
35. Kim B, Kim S, Park J (2012) A school bus scheduling problem. Eur J Oper Res 218:577–585
36. Langevin A, Mbaraga P, Campbell JF (1996) Continuous approximation models in freight distribution: an overview. Transp Res B-Methodol 30(3):163–188
37. Larson RC, Odoni AR (1981) Urban operations research. Prentice-Hall, Englewood Cliffs
38. Li L, Fu Z (2002) The school bus routing problem: a case study. J Oper Res Soc 53:552–558
39. Li X, Quadrifoglio L (2010) Feeder transit services: choosing between fixed and demand responsive policy. Transp Res C-Emerg 18:770–780
40. Metcalf J (2012) (J. Campbell, Interviewer)
41. Miranda PA, Gonzalea-Ramirez RG, Smith NR (2011) Districting and customer clustering within supply chain planning: a review of modeling and solution approaches. In: Renko S (ed) Supply chain management—new perspectives. InTech, Rijeka, Croatia, pp 737–770
42. Missouri Department of Elementary & Secondary Education (2012) http://dese.mo.gov/documents/Snapshot-of-Public-Education.pdf. Accessed 12 Apr 2012
43. National Center for Safe Routes to School; Pedestrian and Bicycle Information Center (2010) Selecting school bus stop locations: a guide for school transportation professionals. National Highway Traffic Safety Administration, July 2010
44. Nourbakhsh SM, Ouyang Y (2012) A structured flexible transit system for low demand areas. Transp Res B-Methodol 46:204–216
45. Novaes AG, Graciolli OD (1999) Designing multi-vehicle delivery tours in a grid cell format. Eur J Oper Res 119:613–634
46. Pacheco J, Marti R (2006) Tabu search for a multi-objective routing problem. J Oper Res Soc 57:29–37
47. Park J, Kim B (2010) The school bus routing problem: a review. Eur J Oper Res 202:311–319
48. Park J, Tae H, Kim B (2012) A post-improvement procedure for the mixed load school bus routing problem. Eur J Oper Res 217:204–213
49. Park J, Tae H, Kim B (2012) Corrigendum to "Post-improvement procedure for the mixed load school bus routing problem". Eur J Oper Res 217:204–213
50. Park J, Tae H, Kim B (2013) Corrigendum to "Post-improvement procedure for the mixed load school bus routing problem". Eur J Oper Res 226:661–662
51. Prasetyo D, Muhamad J, Fauzi R (2011) Supporting needy student in transportation: a population based school bus routing in spatial environment. International Conference on Social Science and Humanity. IACSIT Press, Singapore, pp 48–53
52. Quadrifoglio L, Li X (2009) A methodology to derive the critical demand density for designing and operating feeder transit services. Transp Res B-Methodol 43:922–935
53. Quadrifoglio L, Hall RW, Dessouky MM (2006) Performance and design of mobility allowance shuttle transit services: bounds on the maximum longitudinal velocity. Transp Sci 40(3):351–363

54. Russell R, Morrel R (1986) Routing special-education school buses. Interfaces 16:56–64
55. Safe Routes to School National Partnership (2013) http://www.saferoutespartnership.org/sites/default/files/pdf/What-is-SRST-factsheet-REVISED-06-14-11-w-footnotes.pdf. Accessed 25 Oct 2013
56. Sankaran JK, Wood L (2007) The relative impact of consignee behaviour and road traffic congestion on distribution costs. Transp Res B-Methodol 41:1003–1049
57. Spada M, Bierlairs M, Liebling TM (2005) Decision-aiding methodology for the school bus routing and scheduling problem. Transp Sci 39:477–490
58. Szplett DB (1984) Approximate procedures for planning public transit systems: a review and some examples. J Adv Transp 18(3):245–257
59. Thangiah SR, Nygard KE (1992) School bus routing using genetic algorithms. In: SPIE conference on applications of artificial intelligence X: knowledge-based systems. Orlando, FL, pp 387–398
60. Thangiah SR, Fergnay A, Wilson B, Pitluga A, Mennell W (2008) School bus routing in rural school districts. In: Hickman M, Mirchandani P, Voss S (eds) Computer-aided systems in public transport. Springer, Heidelberg, Germany, pp 209–232
61. Tsao Y-C, Lu J-C (2012) A supply chain network design considering transportation cost discounts. Transp Res E-Log 48:401–414
62. Turkensteen M, Klose A (2012) Demand dispersion and logistics costs in one-to-many distribution systems. Eur J Oper Res 223:499–507
63. van Nes R, Bovy PH (2000) Importance of objectives in urban transit network design. Transp Res Rec 1735:25–34
64. Vaughan R (1984) Approximate formulas for average distances associated with zones. Transp Sci 18(3):231–244
65. Zhao J, Dessouky M (2008) Service capacity design problems for mobility allowance shuttle transit systems. Transp Res B-Methodol 42:135–146

Part II
Hub Location

Some Numerical Studies for a Complicated Hub Location Problem

J. Fabian Meier and Uwe Clausen

Abstract We consider a complicated hub location problem which includes multi-allocation, different hub sizes and different transport volumes on different week days. Furthermore, we consider transport costs per vehicle and not per volume which transforms the cost function into a step function and makes the problem numerically very hard. In our previous work we developed a heuristic approach which we now want to compare to CPLEX results for general and simplified models.

1 Introduction

Hub location problems have become classic challenges in the area of discrete optimization. The original problems, as they are very well described in [1], use a graph of depots which are connected by transport arcs. The task is to transport a given set of shipments from their sources to their sinks in a cost-optimal way. For that, some depots are equipped as *hubs*; then one assigns to every shipment a path from its source to its sink using only hubs in between. The total cost is the sum of the transport costs (for every shipment on every arc it uses) and the costs for the hubs (some problems require a fixed number of hubs p which is equivalent to assigning zero cost to the first p hubs and infinite costs to the following ones).

The main idea of hub location problems is *economies of scale*: Bundling shipments usually decreases the unit transport costs and may hence be beneficial even if it requires detours and costly facilities. Classic hub location problems usually assume

Supported by Deutsche Forschungsgemeinschaft, Project *Lenkung des Gueterflusses in durch Gateways gekoppelten Logistik-Service-Netzwerken mittels quadratischer Optimierung.*

J.F. Meier (✉)
Institut Fuer Transportlogistik, TU Dortmund, Leonhard-Euler-Str. 2, 44227 Dortmund, Germany
e-mail: meier@itl.tu-dortmund.de

U. Clausen
Institut Fuer Transportlogistik, Fraunhofer-Institut Fuer Materialfluss Und Logistik, Joseph-von-Fraunhofer-Str. 2-4, 44227 Dortmund, Germany
e-mail: clausen@itl.tu-dortmund.de

H.-J. Sebastian et al. (eds.), *Quantitative Approaches in Logistics and Supply Chain Management*, Lecture Notes in Logistics,
DOI 10.1007/978-3-319-12856-6_2

fixed unit costs for each transport arc; to simulate economies of scale they reduce the unit costs on hub-hub-connections by a factor α because "usually" these arcs carry more overall weight. This simplification increases the solvability but also limits the applicability of the model.

We investigate the problem from the point of view of a less-than-truckload network planner. A large number of small shipments has to be transported from their source depots to their sink depots. A vehicle can transport many of these shipments, so that it is advantageous to bundle shipments for transport: Instead of direct transport we establish hubs as transhipment points. We ask the strategic questions:

- Where should hubs be established?
- What transhipment capacities should be assigned to them?

To give a cost-efficient answer to such questions, we have to balance the strategic costs of establishing transhipment capacities with the prospective tactical/operational costs of the transport. Our model incorporates the following challenges:

- *Truck-based transport costs*. If we send a truck from A to B, the resulting cost depends on general vehicle costs, driver salary and fuel consumption. The filling quota of the truck has little influence on the fuel consumption and nearly no influence on the other terms. Thus we obtain a good approximation of the real costs if we measure transport costs "by vehicle" instead of "by volume" [2]. On each connection $A \to B$, the cost per volume becomes a step function. Such vehicle based costs were already considered in the mathematical models of [8, 9].
- *Multi-allocation*. Every shipment can be independently routed, so that each depot can be connected to many others. If it turns out to be cheaper to have some direct transports, this is also possible.
- *A weekday-based schedule*. We consider a European network where the travel time between two depots/hubs is one to four days. As our shipping volumes depend on the day of the week, we use a cyclic model with five time slices to represent the working days.
- *Variable hub sizes*. We assign a transhipment to every possible hub. It can be chosen on a continuous scale. Hubs of different capacities were considered by [7], but our weekly schedule adds an additional flavour: Strategic decisions have to be equal for every day of the week, i.e. the transhipment capacity of a hub is the same on every week day.
- *Buffering*. We want to analyse the effect of buffering, i.e. the possibility of storing a shipment in a hub for a day to get a cheaper transport on the next day. Therefore, we consider a scenario with a separate buffering capacity for every hub, which can also be chosen on a continuous scale at the strategic level. Buffering is considered as a "transport in time" lasting one day. In principle, buffering actions can be chained, but as our real world instances have strict transport time limits this is usually not possible.

Section 2 will define and discuss mathematical models for the strategic planning problem that was just outlined. These have a huge number of binary variables, so that we define restricted models in Sect. 3 that are easier to solve. Section 4 briefly describes our heuristic approach which is detailed in [3]. In Sect. 5, we state and discuss the numerical results of both the MIP approaches and the heuristic approach. A short conclusion completes the paper.

2 The Mathematical Models

We start with a given set of D of *depots*. The depots are all interconnected by *transport connections*. A transport connection can be used by shipments to get from one depot to another; it has a *travel time* (in days) and *cost factor* which is the cost per vehicle on this connection. We consider a homogeneous fleet as one usually uses the maximal allowed truck size on European connections. The situation can be depicted by a directed graph: The nodes are formed by the (depot, weekday) pairs, and a transport connection from depot A to depot B which needs n days connects (depotA, d) with (depotB, $d + n \pmod 5$) for each weekday d (shown in Fig. 1).

Furthermore we have a large number of shipments. These shipments all have a source depot, a sink depot, a volume and a maximal travel time. The routes can use every depot for transhipment or buffering which is equipped with the appropriate hub capacity. The transhipment capacities of the hubs have to be chosen large enough to work for *every* weekday, i.e. they need to handle the maximal transhipment that happens throughout the week. A route can consist of arbitrary many steps as long as the maximal travel time is not exceeded.

There are two general approaches to model the routing of the shipments: A *route-based* or a *flow-based* view.

In the route-based view, each possible route for a shipment is represented by a binary variable from which exactly one has the value 1. Without any restrictions on

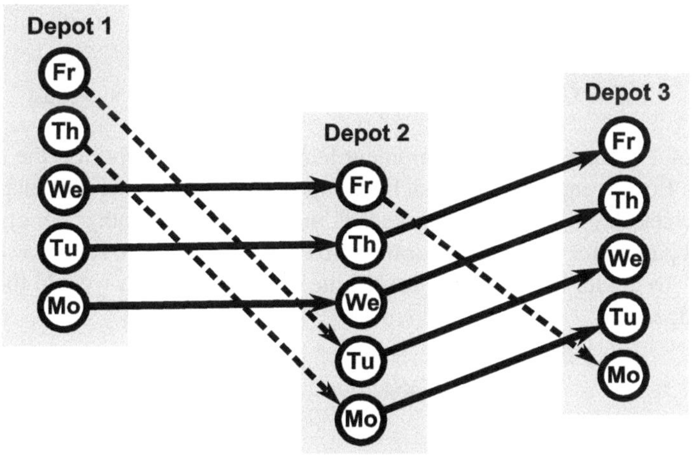

Fig. 1 The time expanded network: each arc represents a movement in space and (cyclic) time

the possible paths, the number of variables is exponential in the number of edges of the graph. Hence this kind of model can only be sensibly applied when we have strong restrictions on the number of possible paths (like in our paper [5] or in classical hub location problems [1], where at most three edges per path were allowed) or apply an approach like column generation. As column generation failed to give good results in a simpler model with truck-based costs [6], we will not use this approach in the general case, but we will come back to it in the next section to build restricted models.

The flow-based view is inspired by the multi-commodity flow problem, but with the major difference that shipments cannot be split, so that we need a binary variable for each (shipment, edge) pair, stating if the edge is used by the shipment. As the number of commodities and the number of edges are each of order $(\#\text{depots})^2 \cdot (\#\text{days per week})$, we get approximately $(\#\text{depots})^4 \cdot (\#\text{days per week})^2$ binary variables. Due to the maximal travel time of each shipment, some of these variables can be set to zero in preprocessing.

Firstly, we will construct a mathematical model without shipment buffering, then we will add this feature later. Let D be the set of depots and $W = \{0, 1, 2, 3, 4\}$ be the set of week days. Furthermore we have a set Q of shipments. For every $q \in Q$ we denote by $q_{\text{so}} \in D$, $q_{\text{si}} \in D$, $q_{\text{day}} \in W$, am_q and time_q the source, sink, starting day, amount and maximal travel time respectively.

The most important variable is the binary flow variable f_{qabw}. It states whether the shipment $q \in Q$ uses the arc $a \to b$, $a \neq b$ with starting day w. To form a flow it has to fulfill the following three conditions (tr_{ab} is the number of days to travel from a to b):

$$\sum_{d \in D, d \neq q_{\text{so}}} f_{q q_{\text{so}} d q_{\text{day}}} = 1 \qquad q \in Q \tag{1}$$

$$\sum_{d \in D, d \neq q_{\text{si}}, w \in W} f_{q d q_{\text{si}} w} = 1 \qquad q \in Q \tag{2}$$

$$\sum_{a \in D, a \neq d} f_{qad(w - \text{tr}_{ad} \bmod 5)} = \sum_{b \in D, b \neq d} f_{qdbw}$$
$$q \in Q, d \in D, w \in W, \quad d \neq q_{\text{si}}, \; d = q_{\text{so}} \Rightarrow w \neq q_{\text{day}} \tag{3}$$

Equation (1) states that each shipment leaves its source depot on the respective day, while Eq. (2) indicates that each shipment has to reach the sink depot (on an arbitrary week day). Equation (3) matches the flows for any other depot and time. As the last parameter of f stands for the starting day, we have to reduce w on the left hand side. To model the maximal travel time of each shipment we add the times of all used edges (4):

$$\sum_{a \neq b \in D, w \in W} f_{qabw} \cdot \text{tr}_{ab} \leq \text{time}_q \qquad q \in Q \tag{4}$$

Let us introduce some auxiliary variables: We define t_{abw} to be the transport volume on the edge $a \to b$ starting on day w, v_{abw} the number of vehicles on that edge, u_{dw} the total transhipment at depot d on day w and $u_d^{\max}$ the maximum over w of u_{dw}. These variables are defined by the Eqs. (5–8) (size is the size of a vehicle in units of volume):

$$t_{abw} = \sum_{q \in Q} \mathrm{am}_q \cdot f_{qabw} \qquad a \neq b \in D, w \in W \tag{5}$$

$$v_{abw} \cdot \mathrm{size} \geq t_{abw} \qquad a \neq b \in D, w \in W \tag{6}$$

$$u_{dw} = \sum_{b \in D} t_{dbw} - \sum_{q \in Q: q_{\mathrm{so}}=d, q_{\mathrm{day}}=w} \mathrm{am}_q \qquad d \in D, w \in W \tag{7}$$

$$u_d^{\max} \geq u_{dw} \qquad d \in D, w \in W \tag{8}$$

Using the parameters $\mathrm{truckcost}_{ab}$ for the costs of using a truck on connection $a \to b$ and $\mathrm{transcost}_d$ for the strategic costs of having transhipment capacity, we can state the objective function as:

$$\sum_{a,b \in D, w \in W} v_{abw} \cdot \mathrm{truckcost}_{ab} + \sum_{d \in D} u_d^{\max} \cdot \mathrm{transcost}_d \tag{9}$$

Let us call this problem *Multi-Allocation Weekday Scheduled Strategic Planning Problem* MAWSSPP. To get the buffering version BMAWSSPP, we need to introduce the possibility to store a commodity in a hub for a day. For this, we use the flow variables f_{qddw} which describe a "transport" from d to itself lasting one day. The buffering is also a strategic cost which is charged similarly to the transshipment cost (but with the factor buffcost). For that b_{dw} and $b_d^{\max}$ are analogously defined to u_{dw} and $u_d^{\max}$. We write:

$$b_{dw} = \sum_{q \in Q} f_{qddw} \cdot \mathrm{am}_q \qquad d \in D, w \in W \tag{10}$$

$$b_d^{\max} \geq b_{dw} \qquad d \in D, w \in W \tag{11}$$

The constraints (1–5) are transformed to:

$$\sum_{d \in D} f_{q q_{\mathrm{so}} d q_{\mathrm{day}}} = 1 \qquad q \in Q \tag{12}$$

$$\sum_{d \in D, w \in W} f_{q d q_{\mathrm{si}} w} = 1 \qquad q \in Q \tag{13}$$

$$\sum_{a \in D} f_{qad(w - \mathrm{tr}_{ad} \bmod 5)} = \sum_{b \in D} f_{qdbw}$$

$$q \in Q, d \in D, w \in W, \quad d \neq q_{\text{si}}, \ d = q_{\text{so}} \Rightarrow w \neq q_{\text{day}} \tag{14}$$

$$\sum_{a \neq b \in D, w \in W} f_{qabw} \cdot \text{tr}_{ab} + \sum_{d \in D, w \in W} f_{qddw} \leq \text{time}_q \qquad q \in Q \tag{15}$$

$$t_{abw} = \sum_{q \in Q} \text{am}_q \cdot f_{qabw} \qquad a, b \in D, w \in W \tag{16}$$

These five equations only differ from their counterparts by usage of buffering flows f_{qaaw}. For (15) we added a term adding one day for every buffering. We note that the (following) Eqs. (17) and (19) are unchanged, while (18) gets an additional term: Without it, buffered shipments would be charged twice for transhipment, but we chose only to charge incoming shipments.

$$v_{abw} \cdot \text{size} \geq t_{abw} \qquad a \neq b \in D, w \in W \tag{17}$$

$$u_{dw} = \sum_{b \in D} t_{dbw} - \sum_{q \in Q: q_{\text{so}}=d, q_{\text{day}}=w} \text{am}_q - b_{d(w-1 \bmod 5)} \qquad d \in D, w \in W \tag{18}$$

$$u_d^{\max} \geq u_{dw} \qquad d \in D, w \in W \tag{19}$$

The cost function is extended by an extra term:

$$\sum_{a,b \in D, w \in W} v_{abw} \cdot \text{truckcost}_{ab} + \sum_{d \in D} u_d^{\max} \cdot \text{transcost}_d + \sum_{d \in D} b_d^{\max} \cdot \text{buffcost}_d \tag{20}$$

Let us note that our modelling of the buffering feature includes the possibility of chaining buffering edges which means buffering a shipment for more than one day. We see no theoretic reasons for stronger constraints, but in practice the maximal transport time restrictions often exclude long buffering.

One can improve the solvability of the problem by discarding some variables in preprocessing. The largest number of eliminated variables can normally be achieved by the following argument:

$$f_{qabw} = 1 \quad \Rightarrow \quad \text{tr}_{q_{\text{so}}a} + \text{tr}_{ab} + \text{tr}_{bq_{\text{si}}} \leq \text{time}_q, \tag{21}$$

if we make the reasonable assumption that transport times fulfill the triangle inequality. Variables not fulfilling condition (21) can hence be eliminated from the equations.

3 Restricted Mathematical Models

A way to improve solvability of the MAWSSPP is to drastically reduce the number of involved binary variables. We want to define restricted models whose solutions are still valid solutions for MAWSSPP (and hence also for BMAWSSPP). We consider two approaches:

1. We consider the same transport plan for every day (SAMEDAY). By this, the number of routes that have to be assigned is reduced by a factor of five. Furthermore, it reflects the reality in many non-automatized settings.
2. We allow at most one hub on each route (ONEHUB). This leads to a massive reduction in the number of binary variables (detailed below).

Let us first discuss the elimination of weekdays. Until now, we assumed one shipment for every (source depot, sink depot, week day) triple, possibly of size zero. Now we define a "super shipment" for every pair (source depot, sink depot) which has the maximal size of all five attached shipments. We call the set of super shipments Q^s and furthermore reuse the variables f, v, t and u, which are now time independent (we can thus dispense with $u^{\max}$). Solving the routing problem for these super shipments (with quintupled costs) automatically gives a solution for the original problem. The model now looks like this:

$$\sum_{d \in D, d \neq q_{\text{so}}} f_{q q_{\text{so}} d} = 1 \quad q \in Q^s \tag{22}$$

$$\sum_{d \in D, d \neq q_{\text{si}}} f_{q d q_{\text{si}}} = 1 \quad q \in Q^s \tag{23}$$

$$\sum_{a \in D, a \neq d} f_{qad} = \sum_{b \in D, b \neq d} f_{qdb} \quad d \in D, d \neq q_{\text{si}}, d \neq q_{\text{so}} \tag{24}$$

$$\sum_{a \neq b \in D} f_{qab} \cdot \text{tr}_{ab} \leq \text{time}_q \quad q \in Q^s \tag{25}$$

$$t_{ab} = \sum_{q \in Q^s} \text{am}_q f_{qab} \quad a \neq b \in D \tag{26}$$

$$v_{ab} \cdot \text{size} \geq t_{ab} \quad a \neq b \in D \tag{27}$$

$$u_d = \sum_{b \in D} t_{db} - \sum_{q \in Q^s : q_{\text{so}} = d} \text{am}_d \quad d \in D \tag{28}$$

$$\min \; 5 \cdot \left(\sum_{a,b \in D} v_{ab} \cdot \text{truckcost}_{ab} + \sum_{d \in D} u_d \cdot \text{transcost}_d \right) \tag{29}$$

For the ONEHUB model, we opted for a route-based-approach, because we can now easily describe the routing of a shipment by giving the intermediate hub d as

r_{qd}. A direct routing can be represented by $d = q_{\text{si}}$ or $d = q_{\text{so}}$. In this way we reduce the number of binary variables from about (#days per week)^2 · (#depots)^4 to approximately (#days per week) · (#depots)^3.

We keep all the other variables except for f. Hence, we can leave the objective function unchanged. Essentially, we have to make four changes:

- Delete the flow conditions (1–4).
- The transport volume t_{abw} is now calculated as:

$$t_{abw} = \sum_{\substack{q \in Q, \\ q_{\text{so}}=a, \\ q_{\text{day}}=w}} r_{qb} \cdot \text{am}_q + \sum_{\substack{q \in Q, \\ q_{\text{si}}=b, \\ q_{\text{day}}=w-\text{tr}_{da}}} r_{qa} \cdot \text{am}_q \quad a \neq b \in D, w \in W \tag{30}$$

- We have to ensure that for every commodity exactly one route is chosen:

$$\sum_{d \in D} r_{qd} = 1 \quad q \in Q \tag{31}$$

- The maximal travel time constraint has to be rewritten:

$$\sum_{d \in D} r_{qd} \cdot \left(\text{tr}_{q_{\text{so}}d} + \text{tr}_{dq_{\text{si}}}\right) \leq \text{time}_q \quad q \in Q \tag{32}$$

In our paper [4, Sect. 4] we considered additional inequalities to strengthen the formulation, which were not very successful in the two-hub-case. Following Martin Baumung's good unpublished results for the one-hub-case, we reconsider them. The idea is that we can calculate the minimal flow from a subset $K \subset D$ to $D \backslash K$ as

$$\text{minflow}\left(K, D\backslash K\right) = \sum_{q \in Q, q_{\text{so}} \in K, q_{\text{si}} \in D \backslash K} \text{am}_q \tag{33}$$

From that we know that the number of vehicles going from K to $D\backslash K$ is at least $\lceil \text{minflow}(K, D\backslash K)/\text{size} \rceil$. Due to the rounding up procedure, this bound is stronger than the original LP bound. To avoid adding huge numbers of inequalities, we consider this only for $|K| = 1$ and $|K| = |D| - 1$. In the first case, we can even consider the outgoing flow of every day separately. In the end, we get:

$$\sum_{b \in D} v_{abw} \geq \left\lceil \left(\sum_{q \in Q: q_{\text{so}}=a, q_{\text{day}}=w} \text{am}_q \right) / \text{size} \right\rceil \quad a \in D, w \in W \tag{34}$$

$$\sum_{a \in D, w \in W} v_{abw} \geq \left\lceil \left(\sum_{q \in Q: q_{\text{si}}=b} \text{am}_q \right) / \text{size} \right\rceil \quad a \in D \tag{35}$$

4 A Short Description of the Heuristic Approach from [3]

In [3] we developed a general modelling language for shipment based transport problems, i.e. problems that consist of a large number of (unsplittable) shipments which have to be transported through a graph. It is based upon the following paradigms:

1. The routes of the shipments are considered as independent variables, while all other variables (like number of trucks, transhipment capacities, buffering capacities, etc.) are calculated from the chosen routes.
2. Each shipment has a set of admissible routes. Constraints that depend on more than one route are modelled in the cost function.
3. A neighbour of a given solution is created by taking a small subset of the shipments and replacing their routes by others. To avoid extremely large neighbourhoods we discard neighbours which fail to fulfill a special local optimality criterion detailed in [3].

The model and the neighbourhood creation scheme allow us to implement a Simulated Annealing algorithm. The numerical results are given in the next section.

5 Numerical Results

We want to compare three different approaches:

1. The full model solved by CPLEX.
2. Heuristic results based upon Sect. 4.
3. The results of CPLEX for the restricted models.

We will use the seven benchmark instances I5, I10, I20, I30, I40, I50 and I60 with the respective number of depots which are based upon data from a large European road freight company. We solved each of it twice: with buffering and without buffering. The results are summarized in Table 1. We used a computer with 3.4 GHz and 16GB RAM for six hours, both for the heuristic and for CPLEX 12.6.0.

Contrary to our expectation, the buffering advantage does not show up in our heuristic results. Especially for the larger instances, we tend to get better results in the non-buffering case. There are practical and numerical reasons for this: Firstly, the option "buffering" increases the size of the search space and so slows down the heuristic. Secondly, the larger instances offer more other possibilities for consolidation so that buffering is not so important.

Comparing heuristic and CPLEX we see that although we drastically reduced the number of variables by preprocessing CPLEX fails for each of the instances over 20 depots. We see that the heuristic works well for the small instances; for the larger ones, we have no comparison.

Hence we solved the restricted models SAMEDAY and ONEHUB with the same solver and computer. The results are shown in Tables 2 and 3. We see that our heuristic outperforms all of them.

Table 1 Results of the heuristic approach compared to the results of CPLEX with preprocessing

Inst	Heur	Heur buf	CPLEX	CPLEX buf
I5	29,206,387	19,221,750	29,206,387	19,221,750
LB			29,206,387	19,221,750
I10	101,029,802	90,112,421	101,029,802	91,972,141
LB			101,029,802	82,633,347
I20	156,016,297	149,190,951	201,129,327	No solution
LB			135,228,841	No solution
I30	324,679,889	347,634,809	No solution	No solution
I40	468,567,791	470,264,568	No solution	No solution
I50	668,569,617	683,945,737	No solution	No solution
I60	978,018,115	977,338,269	No solution	No solution

Every problem is considered with and without the possibility of buffering

Table 2 Results from SAMEDAY compared to the best known results

Inst	SAMEDAY	Lower bound	Best known	Best known with buff
I5	30,171,383	30,171,383	29,206,387	19,221,750
I10	133,023,575	133,023,575	101,029,802	90,112,421
I20	222,652,292	163,991,639	156,016,297	149,190,951
I30	597,737,936	363,056,830	324,679,889	324,679,889
I40	No solution		468,567,791	468,567,791
I50	2,169,759,985	670,424,018	668,569,617	668,569,617
I60	No solution		978,018,115	977,338,269

Table 3 Results from ONEHUB (minimum of the results with and without strengthening inequalities) compared to the best known results

Inst	ONEHUB	Lower bound	Best known	Best known with buff
I5	29,206,387	29,206,387	29,206,387	19,221,750
I10	119,128,292	119,128,292	101,029,802	90,112,421
I20	200,685,015	188,607,303	156,016,297	149,190,951
I30	642,446,816	295,834,506	324,679,889	324,679,889
I40	7,881,411,010	381,556,712	468,567,791	468,567,791
I50	3,225,667,487	526,796,494	668,569,617	668,569,617
I60	4,813,948,039	739,362,497	978,018,115	977,338,269

6 Conclusion

The results show that the realistic hub location problem that we stated is very difficult for standard MIP solvers. This difficulty persists not only if we do preprocessing but also when we drastically reduce the complexity of the model. On the other hand, a heuristic approach based on [3] performs well. Hence we will follow two paths for the further development:

On the one hand, we will improve our heuristic by proper gauging. A heuristic procedure involves a huge number of search parameters which have to be calibrated by statistical methods. On the other hand, we will aim for better lower bounds by solving relaxed hub location problems, preferably with a Benders' decomposition approach.

References

1. Alumur S, Kara BY (2008) Network hub location problems: the state of the art. Eur J Oper Res 190(1):1–21
2. Fleischmann B (2008) Transport- und Tourenplanung. In: Arnold D, Isermann H, Kuhn A, Tempelmeier H, Furmans K (eds) Handbuch logistik, vol 3. Springer, pp 137–152
3. Meier JF (2014) A versatile heuristic approach for generalized hub location problems. Preprint, Provided upon personal request
4. Meier JF, Clausen U (2013) Heuristic strategies for a multi-allocation problem in LTL logistics. In: Operations research proceedings 2012: selected papers of the international annual conference of the German operations research society (GOR). Springer, Germany
5. Meier JF, Clausen U (2013) Strategic planning in LTL logistics—increasing the capacity utilization of trucks. Electron Notes Discret Math 41:37–44
6. Meier JF, Clausen U, Baumann F (2013) A column generation approach for strategic planning in LTL logistics. In: Proceedings of VeRoLog 2013
7. Sender J, Clausen U (2011) Hub location problems with choice of different hub capacities and vehicle types. In: Pahl J, Reiners T, Voss S (eds) Network optimization. Springer, Berlin, pp 535–546
8. Sender J, Clausen U (2013) Heuristics for solving a capacitated multiple allocation hub location problem with application in German wagonload traffic. Electron Notes Discret Math 41:13–20
9. Voll R, Clausen U (2013) Branch-and-price for a European variant of the railroad blocking problem. Electron Notes Discret Math 41:45–52

Part III
Supply Chain Management

Maintenance Enterprise Resource Planning: Information Value Among Supply Chain Elements

Rogers Ascef, Alex Bordetsky and Geraldo Ferrer

Abstract Maintenance Supply Chain (MSC) involves Maintenance, Repair and Overhaul (MRO) organizations and the relationships within and across suppliers and customers. These organizations work with the probability of equipment failure, maintenance and user requirements of spare parts. All of these elements increase uncertainty in this environment. Besides, it is difficult to integrate and process information to maintain good inventory control. This high uncertainty and lack of integration of information cause spare parts inventory excesses and shortages. This research proposes a new model based on information processing theories to connect the lateral elements of the supply chain, increase vertical information and transform the MSC into a system to decrease shortages and excesses of inventory. This research incorporates a simulation to compare the new model with traditional models of inventory control. This study claims that when using the new model with different demands of maintenance, inventory cost is lower than with traditional models of inventory control. The research uses information processing theory as the framework to decrease uncertainty, and consequently decrease excesses and shortages of spare parts in MSC.

1 Introduction

The 2007 United States Census showed that expenses in Repair and Maintenance Service were US $137 billion. In comparison, Aircraft Manufacturing sales were US $84 billion [34]. Fabry and Schmitz-Urban wrote that the maintenance sector in Germany had greater turnover (€ 250 billion) than many other industrial sectors, such

R. Ascef (✉)
Department of Logistics, Brazilian Air Force, Rio de Janeiro, Brazil
e-mail: rogersascef@gmail.com

A. Bordetsky
Department of Information Sciences, Naval Postgraduate School, Monterey, CA, USA
e-mail: abordets@nps.edu

G. Ferrer
Graduate School of Business and Public Policy, Naval Postgraduate School, Monterey, CA, USA
e-mail: gferrer@nps.edu

H.-J. Sebastian et al. (eds.), *Quantitative Approaches in Logistics and Supply Chain Management*, Lecture Notes in Logistics,
DOI 10.1007/978-3-319-12856-6_3

as Vehicle Manufacturing (€ 135 billion) [10]. "American businesses and consumers spend approximately US $1 Trillion every year on assets they already own", a good part of this on maintenance expenses [6, p. 130].

When Pan Am and Eastern Airlines went bankrupt, they held an excess inventory of spare parts of approximately $700 million and $200 million, respectively [19]. In the military environment, a 2009 U.S. Department of Defense (DoD) report stated that nearly 17 % of all items in the inventory were inactive, and they valued approximately US $15 billion [8]. Most of these items had been purchased as spares for maintenance purpose, a problem that illustrates the challenge of managing the Maintenance Supply Chain.

The maintenance environment includes components with stochastic failure rate, different types of failure to be repaired, great numbers of spare parts for repair and long lead-times to perform maintenance and to purchase spare parts. Frequently, maintenance does not incorporate fluctuations in equipment usage, changes in environmental conditions and equipment age [24, p. 18]. The maintenance supply chain elements tend to be disconnected from each other, causing shortages and excesses of materials. All these factors can result in delays and high uncertainty in the maintenance process. High uncertainty and lack of information integration cause excess and shortage of spare parts. This misinformation causes low availability of aircraft, equipment or systems, increasing holding costs.

Some researchers have proposed solutions to mitigate the problem. Ghobbar and Friend studied aircraft companies and found that at least 50 % of companies were not satisfied with their system of inventory control [19]. Newman proposed an MRP model of preventive maintenance [28]. Molinder used simulation to analyze the effects of different sources of uncertainty [27]. Ettkin and Jahnig [9] presented a framework to adapt MRPII to maintenance functions with the benefit of waste reduction. Swanson [33] discussed the use of information-process theory in maintenance management. She conducted a survey in many maintenance, repair and operations (MRO) organizations to show how uncertainty is affected by the use of information systems in maintenance operations. In spite these contributions, the literature still lacks a model that integrates all MRO elements.

This paper seeks to fill this gap. The purpose of this experiment is to test a new integrated model between maintenance supply chain elements to match inventory level with maintenance requirements to decrease inventory cost. This study compares the new model with traditional inventory model of control with different amounts of maintenance demand to inventory costs. This research is important because the result reduces uncertainty and, consequently, decreases cost and increases equipment availability.

This study applies an information processing approach to analyze the information integration between the elements of the maintenance supply chain. It expands on the idea that new information, such as ERP, can increase the capacity of information processing, and consequently can decrease uncertainty and costs. The specific research question addressed in this chapter is:

Does Maintenance Enterprise Resource Planning (MERP) decrease inventory costs compared with the use of traditional inventory models?

This study is divided in five sections: literature review, proposed model, methodology, results and discussions. The proposed model shows how the model integrates the information. The methodology presents the hypothesis and experimental procedure of the research. Finally, the study analyzes and explains the result, and suggests future research.

2 Literature Review

2.1 Information Processing Theory

Frequently, the information about failed components isn't available, maintenance information doesn't integrate across supply divisions, and, the inventory control has to use past information to predict the purchasing material. This entire gap causes high uncertainty in the MSC environment. Galbraith defines "uncertainty as the difference between the amount of information necessary to perform a task and the information already possessed by the company" [17]. He analyzed the relation between uncertainty and information to formulate the information processing theory. His theory claims that "the greater the task uncertainty, the greater the amount of information that must be processed among decision makers during task execution in order to achieve a given level of performance" [16]. He argued that there are two organizational strategies to manage the uncertainty: to reduce the need for information processing or to increase the capacity to process information.

To reduce the need for information requires the creation of slack resources or the existence of self-contained tasks. Moreover, Galbraith indicated that investment in vertical information system and the creation of lateral relations increase the volume to process information. He argued that "the greater the uncertainty, the lower the decision-making and the integration is then maintained by lateral relations" [16].

The concept of this information theory was used in many activities. There are studies in the application of theory to propose structural modification in organizations with vertical analysis and horizontal information systems to increase the information process [5]. Swanson applied the information-processing model to analyze maintenance management [33]. She found that maintenance organizations respond to environmental complexity with the use of computerized maintenance management systems, preventive and predictive maintenance systems, coordination, and increased workforce.

Other research presents a new perception of information sharing within supply chains based on organizational information processing theory. Posey and Bari propose a conceptual model that shows that if information within and across supply chains are more compatible with each other, they can increase information-processing capabilities [29]. Flynn and Flynn explain that some firms found alternatives to processing information by using "management-intensive solutions, rather than technology-intensive solutions" [14, p. 1044].

This study uses the two strategies to coordinate uncertainty in Galbraith information process theory, and compare their efficiency. As the reductionist approach to manage uncertainty, we use the most common model of inventory control: Economic Order Quantity (EOQ). The alternative approach, with increased capacity to process information in the Supply Chain, is the Maintenance Enterprise Resource Planning (MERP).

The two approaches are linked by the ability of the organization to coordinate and process the information. If the firm cannot integrate the information available in multiple departments, if non-routine events are more frequent than the capacity of the firm to process information, or if the technology available cannot increase the information processing capacity of the firm, then the firm must use a reductionist strategy to process information. That is, the firm adopts simple deterministic models for decision making, using basic static information allied to expensive protections, such as inventory buffers, to support the organization in the face of uncertainty.

On the other hand, if the firm can integrate lateral and vertical information within and across organizations, if the firm has low decision-making processing time, and if the firm can integrate the elements of supply chain, then the MERP model can increase the capacity of information processing and decrease the uncertainty in this environment, resulting in lower inventory costs and more responsiveness to any external or internal change. An application of the Galbraith theory with the supply chain model of research is represented in Fig. 1.

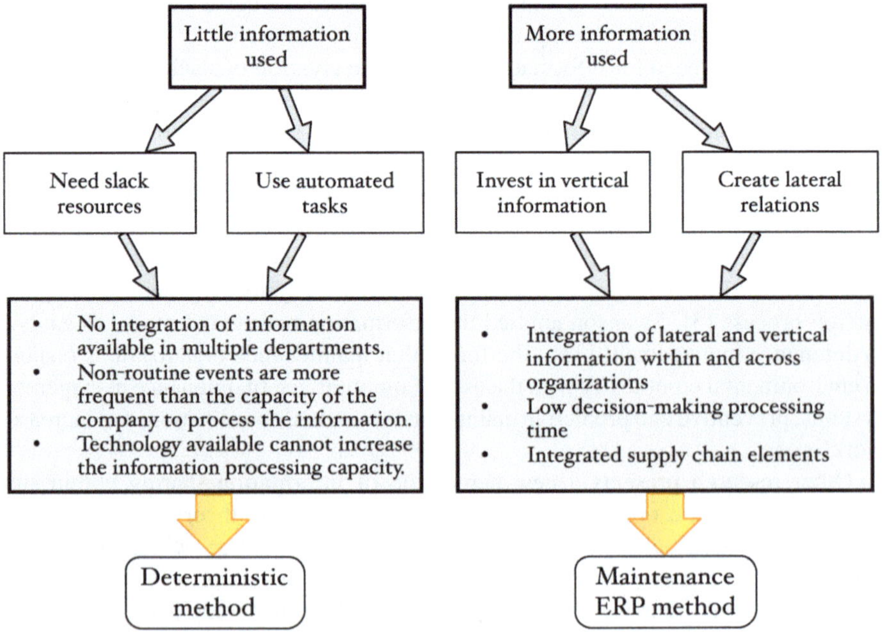

Fig. 1 A supply chain application of galbraith strategies

2.2 Enterprise Resources Planning (ERP)

Vollmann et al. [35] presented two interesting definitions about ERP. For the information technology community, ERP is a term that integrates the application program in finance, manufacturing, logistics, sales and marketing, human resources, and the other functions in an organization. From the manager's viewpoint, ERP represents a comprehensive software approach to support decisions concurrent with planning and controlling the business [35]. ERP seeks to integrate information of the organization through best practice functionality and system interoperability with common databases and interfaces [26].

ERP is an offshoot of the tool Material Requirement Planning (MRP). MRP's function is to prepare a master production schedule (MPS) and a list of materials required for the production process. This technique was developed in 1960 and became more accessible with the development of computers that could process the large database that it requires. Subsequently, this technique evolved into the tool known as Manufacturing Resource Planning (MRP II), which expanded the benefit to the incorporate manufacturing planning beyond materials acquisition. The new technology required more computing power while more integrated decision-making was achieved. ERP is an extension of MRP II that seeks to integrate information and processes across the companies in the supply chain, using electronic data interface (EDI). Interested readers are encouraged to read more in [35].

The proposed model uses ERP techniques to reduce uncertainty in the maintenance supply chain. Ghobbar and Friend [19] surveyed 287 aircraft companies (96 airline operators and 56 maintenance service organizations) to find how they determined reorder point systems for their parts and components for operation and maintenance. They found that 66 % of the maintenance organizations and 57 % of airline operator organizations did not use MRP, "were aware of MRP but had neither used nor investigated it further." The results showed that more than 50 % of companies were not satisfied with their inventory management system [19].

Newman [28] argued that MRP could be used for Preventive Maintenance Requirement Planning where its use could have multiple benefits: part consumption could be tracked and maintenance personnel could be better used. His model showed some aspect for integrating Maintenance Schedule with Supply Chain Management.

Molinder [27] studied how an MRP system was affected by stochastic demand and lead times. He used a "simulation with the objective of analyzing the effects of different sources of uncertainty in MRP systems". He found that high variability had a strong effect on the level of safety stock and safety lead-time required. An adaptation of MRP to maintenance had predicted this uncertainty.

Bojanowski [4] developed a variant of MRP, the Service Requirement Planning (SRP), to prioritize routine mechanical inspection and machine maintenance sequences. Ettkin and Jahnig [9] presented a framework for adapting MRPII to maintenance function for waste reduction. They thought that this model could be used successfully in maintenance management because of the similarities between manufacturing and maintenance processes.

Wemmerlov and Whybark [36] showed different approaches to choose lot size using MRP, and compare a number of alternatives such as Economic Order Quantities (EOQ), Periodic Order Quantities (POQ), Part Period Balancing (PPB), and Wagner-Within Algorithm (WW). Wemmerlov and Whybark [36] demonstrated with no uncertainty, the best result was Wagner-Within Algorithm, but with great computational cost. Under demand uncertainty the inventory cost is 0.19 % higher with EOQ than with WW, and PPB is 0.67 % lower than the WW model. Therefore, all three models can produce good solutions. Under uncertainty, the inventory cost has no difference, "EOQ rule carries with it its own safety stock" [36, p. 16] .

Silver et al. [31] did an experiment with lot sizing for individual items with time-varying demand. They add the Silver-Meal Heuristic (SM) that has similar result with Wagner-Within Algorithm to compare the cost with the other models. They conclude that SM and WW have better cost than the others models [31, pp. 198–218]. Gaither [15] complement with other experiment that include Gaither model. The experiment shows the performance of the models that can be used as guidelines for MRP systems.

Whybark and Williams [37] studied the use of safety stock and safety lead-time in MRP in response to four types of demand uncertainty: demand timing and quantity, and supply timing and quantity.

There is some confusion about remanufactured and maintenance management. The concepts are different, and so is their management; "Remanufactured process is an industrial process in which worn-out products are restored to like-new condition [30, p. 295]". Remanufacturing implies equipment disassembly and complete recovery. "It requires the repair or replacement of worn out or obsolete components and modules" [11, p. 87]. Generally inoperable units are disassembled, cleaned, repaired, and placed in inventory to assemble a new unit. On the other hand, "Maintenance constitute a series of actions necessary to restore or retain an item in an effective operational state" [3, p. 1]. Maintenance Management is the planning and execution of scheduled and unscheduled maintenance to maintain the availability of equipment. Remanufacturing may be considered a type of maintenance.

There are studies evaluating MRP for remanufactured industries such as [7], which proposes a new MRP that calculates the number of units produced each period and the number of components needed to assemble the products [7]. Ferrer and Whybark [11, 12] presents the "first fully integrated material planning system to facilitate the management of remanufacturing facility" [12]. Other researchers seek to find the optimal number of used products, or "cores", to procure and disassemble and the optimal quantities of new parts to procure [18].

So, there are many studies that apply MRP with environmental uncertainty, many examples of MRP's use in a variety of industry sectors, and new MRP's use in the remanufacturing sector. But there are few studies of MRP's use in the maintenance sector; a few models only mention the possibility. This research fills this gap and presents a model that connects the elements of maintenance supply chain.

3 Maintenance Enterprise Resource Planning (MERP)

3.1 The Difference Between Manufacturing and Maintenance Organizations

Why not use traditional MRP/ERP in the MSC since it is used a lot in the manufacturing supply chain? First of all, both environments present uncertainty but the maintenance environment has uncertainty practically in all levels of planning. Cohen affirm that "the majority of existing ERP software programs don't have the capability to manage complex service supply chain scenarios" [6] and Maintenance Supply Chain is one of these scenarios.

The demand of manufacturing supply chain is predictable. On the other hand, MSC is unpredictable because many services are trigged when failure occurs. Even scheduled maintenance is difficult to forecast. Because of the dynamics of MSC environment inventory management uses to pre-position resource to decrease the uncertainty. Manufacturing supply chain tries to maximize velocity of resource. The performance metric in manufacturing supply chain is the fill rate. For MSC, it is availability of equipment [6, pp. 132–133].

To manage MSC, the managers have to work with client information about the equipment as well as failures, operations, utilization forecast. Many times, they cannot forecast when failures will happen. And when it happens, maintenance shops don't know the material that they will use to fix the failure. The material that is used in maintenance is disconnected to production, so uncertainty is present in many processes.

For the manufacturing supply chain, the demand is also challenging, but they know the material to assemble the system and know the material supplier. Lead-time of the supplier may also be varied, but MSC has a lot of variability because many items are discontinued and difficult to purchase.

Sometimes, the maintenance supply chain may use some concepts of the remanufacturing supply chain such as the overhaul of the equipment, but the management of failure, corrective and preventive maintenance, availability of equipment are unique to the maintenance supply chain.

Although there are similarities among manufacturing industries such as the traditional manufacturing process (shop floor scheduling and assembly, e.g.) [18], both involve suppliers, plants and customers. There is, however, significant difference according Table 1.

The different characteristics of the Maintenance Supply Chain show that there is the need to develop a specific planning and control system in this environment. The idea is to adapt the elements of ERP to develop a specific model for the Maintenance Supply Chain.

Table 1 Characteristics of manufacturing supply chain versus maintenance supply chain

	Maintenance supply chain	Manufacturing supply chain
Process [18]	It requires special operational processes and skills, such as disassembly, inspection, testing, and repair	Manufacture follows a logic sequence of production
Time response [6, pp. 131–132]	ASAP (same day or next day)	Standard, can be scheduled
Routing [30]	Probabilistic time and occurrence of maintenance task	Manufacturing task is predictive and assembled with logical form
Inventory management [30]	High level of uncertainty inherent in the maintenance process and unique in corrective maintenance	Fixed material quantity to attend to final product assembly
Bill of material	Probabilistic with no fixed material and quantity	Fixed quantity
Nature of demand [6, pp. 131–132]	Always unpredictable, sporadic	Predictable, can be forecast
Lead time	Uncertain because items can be obsolete, or are no longer manufactured. Unknown suppliers	Suppliers known. Agreements and contracts are done more predictably
Number of SKU [6, pp. 131–132]	High	Limited

3.2 Independent and Dependent Demand

The Maintenance Enterprise Resource Planning—MERP model seeks to connect the elements of MSC and decrease the degree of separation among the elements of supply chain. When these elements are connected, a new collaboration network is formed. These environments allow availability of information, decreasing delay and uncertainty and increasing timely response.

The traditional inventory control system works with the assumption that all items are independent in demand, meaning that the demand for an item is independent of other items. Traditional inventory control for this model is the Economic Order Cost (EOQ) model, Production Order Quantity and Quantity Discount Model [21, pp. 489–490].

Traditional MRP works with assumption that there are independent demand items and dependent demand items. Independent demand items are end-product items in manufacturing, such as an aircraft or engine [35, p. 134]. Dependent demand means that the demand for one item is related to the demand for another item. The items to assemble the aircraft have dependent demand [21, pp. 562–563].

MERP model uses the assumption that maintenance is an independent demand. Scheduled and unscheduled maintenance is performed in aircrafts, engines, generators and landing gears are considered independents events. Dependent demand items are the spare parts that are used to do the maintenance.

Corrective maintenance includes all unscheduled maintenance actions, as a result of system/product failure, to restore the system to a specified condition. Unscheduled Maintenance may be measured in terms of frequency or elapsed time. Preventive maintenance includes all scheduled maintenance actions performed to retain a system or product in a specified operational condition [24, p. 4.18]. It covers periodic inspections, critical-item replacement, periodic calibration, and the like. Preventive maintenance may be measured in terms of frequency or elapsed time. Many items use-time between overhaul (TBO) or a scheduled program of maintenance (e.g., cars with maintenance programming of miles driven; aircraft with maintenance programing of hours flown) [3, pp. 16–17].

3.3 MERP Description

MERP has three modules that are responsible to integrate and process the information within and across the organization, these modules compose the Planning System. The first module is Maintenance and Operation Planning (MOP) that calculates a long time corrective and preventive maintenance forecast based in client information (e.g., failure rate, equipment use). MOP calculates per-year, the quantity of maintenance and the budge. If this scenario is feasible, the information is transferred to MMPS; if not, new scenario is calculated.

If the scenario is approved, Master Maintenance Planning Schedule (MMPS) calculates the quantity of maintenance per period. To calculate, the MMPS takes information on the items in stock and in production. Afterwards, this function produces the quantity of Work Order that has to be opened. The information of work order is then transferred to Maintenance Material Requirement Planning (MMRP). Based on a bill of maintenance that is dynamic, updates are made to the work order and the system then calculates the quantity of material that is needed to do the maintenance. Afterwards, the MMRP takes information of stock, acquisition, transportation and lead-time, and calculates the quantity that has to be purchased. If this scenario is feasible, the information is transferred to CMMS and PMS; if not, a new scenario is calculated. The representation of MERP is in Fig. 2. The correspondence between some modules of MRPII and MERP is in Table 2.

3.4 MERP-System Integration and Operation

This section explains the main tasks of each system and how the information are integrated and processed. The explanation is based on Fig. 2.

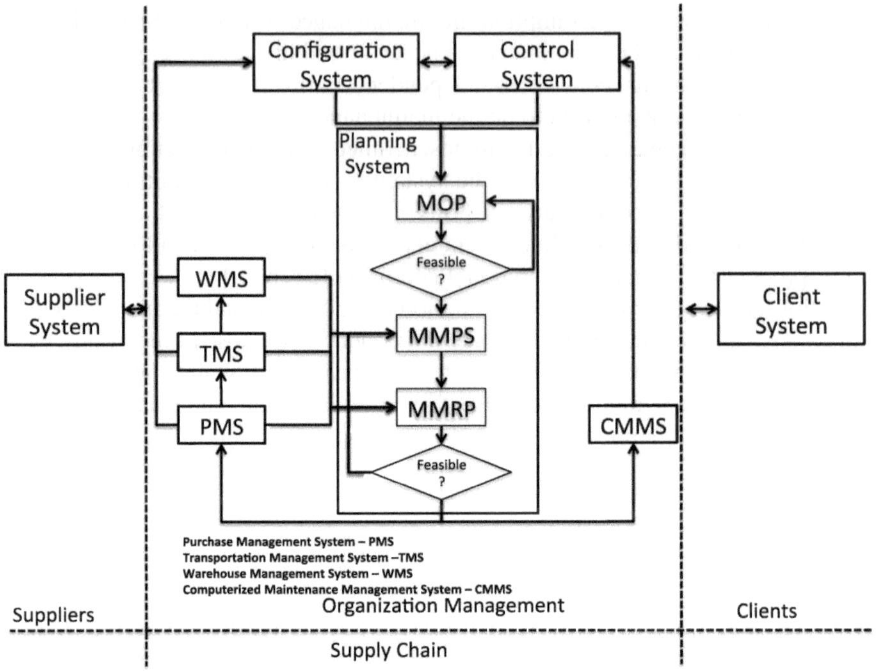

Fig. 2 MERP representation

Table 2 Correspondence between MRP and MERP modules

Traditional MRPII	MERP
SOP—Sales and Operation Planning	MOP—Maintenance and Operating Planning
MPS—Master Production Schedule	MMPS—Maintenance Master Planning Schedule
MRP—Material Requirement Planning	MMRP—Maintenance Material Requirement Planning

- **Configuration System**
  The main tasks of this function are:
  - Basic Information: this function is responsible for registering the initial information of the system and its components as part number, NSN, unit of issue, and price.
  - Primary Configuration: this function is responsible for registering the basic configuration of the reparable items of the system. The system can be composed of many reparable items. This function assembles the structure of system with quantity and position. Example: One car has two batteries, two air conditioners, or, an airplane has two engines, two generators. An engine of an aircraft has two fuel pumps.

  - Maintenance Configuration: this function permits the registration of the type of maintenance that the system and its repairable have. It records the type of maintenance (e.g., Preventive/Predictive Maintenance or Corrective Maintenance), the maintenance cycle, MTBUR, maintenance tasks, tools, man/hour and material that is need to do the maintenance.

  Information shared:

  - With information about maintenance performed in the organization and at the client, the system updates the information about configuration, and maintenance to send to Planning System (e.g., MTBUR, TBO, maintenance time, lot size, lead time).

- **Control System**
  The main tasks of this function are:

  - Utilization Control: this function controls the use of equipment and its repairable items in the organization and at the client, such as the system records prediction of the use of equipment.
  - Reliability Control: based on failure and maintenance data and utilization of the item, this function calculates the Mean Time Between Failure—MTBF and Mean Time Between Unscheduled Replacement (MTBUR) of the reparable item. This function sends information to Maintenance Configuration about the MTBUR of the item.

  MTBUR is the probability of remove a reparable and replace some spare in unscheduled maintenance part during a given period under specified operating conditions [3, p. 2,112].

$$MTBUR = \frac{1}{\lambda} \tag{1}$$

  where λ is referred as the remove and replace spare part spare in unscheduled rate.

  - Maintenance Control: this function controls maintenance cost, the maintenance due date, man-hours used, and life cycle cost.

  Information shared:

  - This function sends information about MTBUR and use of equipment (e.g., update MTBUR, forecast of use of equipment, numbers of equipment in use).

- **Purchase Management System—PMS:**
  The main tasks of this function are:

  - This function control and execute the purchases to the organization.

  Information shared:

- This function receives the purchase planning and updates the stages of purchasing processes and delivery time. This function sends information to MMPS and the MMRP algorithm.

- **Transportation Management System—TMS:**
  The main tasks of this function are:
  - This function plan, control the transportation of equipment and spare parts from clients and suppliers.

  Information shared:
  - This function supplies information about transportation of the item. It supplies data to MMPS and MMRP.
- **Warehouse Management System—WMS**
  The main tasks of this function are:
  - This function controls the stock of the warehouses by receiving, picking and shipping the material.

  Information shared:
  - This function controls the stock and gives information about the quantity of material in stock to MOP, MMPS and MMRP.
- **Computerized Maintenance Management System—CMMS:**
  The main tasks of this function are:
  - This function plan and control the execution of maintenance tasks and updates the information about the material and man/hours that are used in Maintenance Configuration.

  Information shared:
  - This function receives the maintenance planning and updates the stages of the maintenance processes and delivery time. This function sends information to configuration system, MMPS and MMRP algorithm.
- **Client System**
  This module connects information between the client and organization management. The communication can use electronic data interchange (EDI), machine to machine (M2M) techniques, or client-server architecture.
  - Item Information: this function is responsible to register the initial information of the equipment, such as the serial number of a part number, manufacture data, or lifetime.
  - Real Configuration Management: this function is responsible for assembly of the actual configuration of the equipment. This function controls when the item was installed or removed from the equipment.

  - Computerized Maintenance Management System (CMMS): this function registers and controls maintenance that is done with the client, and updates the information about the material and man/hours that are used in Maintenance Configuration.
  - Warehouse Management System–WMS: WMS controls the stock with the client, if it is needed, and connects the information about the stock with organization's management.

- **Supplier System**
  This module connects information with suppliers. The communication can use electronic data interchange (EDI), machine-to-machine (M2M) techniques, or client-server architecture. The information about stock, purchase, reliability, and transportation are shared and exchanged in this function.
- **Planning System**
  Planning System is formed by three modules that connect and process information with the others systems.

- Maintenance and Operation Planning (MOP)
  This function calculates the quantity of corrective maintenance (CM) and preventive maintenance (PM) in a long-time period (2–5 years). This function receives information about MTBUR, TBO, Configuration, Utilization Forecast, Preventive and Corrective Maintenance Cost and calculates the quantity of maintenance in a period.
  A generator of an aircraft is used to illustrate the maintenance forecast calculate. This scenario has 300 aircraft; the quantity per assembly (QPA) is 2 generators. The forecast is to fly an average of 75 h per month for each aircraft by year y and y + 1. MTBUR rate is 5,000 h, and the Time Between Overhaul (TBO) is 3,000 h. These parameters calculate an estimation of maintenance per year. The parameters are in the Table 3.
  To calculate the average quantity of maintenance, the parameters are multiplied. The formula is at Table 4. The PM maintenance is the same as the average calculated. For CM, a service level (k) is entered to find the item in the stock. In this example, Poisson distribution is used, but it can use another distribution depending on the item. It was used with 90 % probability to find the item in stock when it was required. The result is at Table 4.
- Master Maintenance Planning Schedule (MMPS)

Table 3 Parameters to calculate the quantity of Corrective and Preventive Maintenance

Year	QPA	# of aircraft	Utilization per month	MTBUR	TBO	Period
				5,000	3,000	
	x	y	h	$\lambda = 1/\text{MTBUR}$	$z = 1/\text{TBO}$	t
y	2	300	75	0.001	0.0003	12
y+1	2	300	75	0.001	0.0003	12

Table 4 MOP—Quantity of corrective and preventive maintenance

Average CM	Average PM	SL(k)	Qtt CM	Qtt PM
$\mu(\text{cm}) = \text{x y h}$ λ t	$\mu(\text{pm}) = \text{x y h z t}$		$p(k;\lambda) = \frac{\lambda^k}{K!}e^{-\lambda}$	
108	180	0.9	121	180
108	180	0.9	121	180

To calculate the quantity of maintenance that a maintenance shop has to do in a period of time, the model sums the quantity of CM and PM, the quantity of maintenance of a specific reparable and decreases the quantity of equipment that it has in stock and work orders.
Basically, to calculate the master maintenance planning, this function takes information from the Configuration System about the average of maintenance time (MT) of PM and CM, lot size (LS) to do the maintenance (if applicable), and safety stock (SS) of the reparable. To illustrate the calculation, the maintenance time is 1 period; safety stock is 0, and lot size is 1.

The elements of MMPS are:

Maintenance Forecast (MF), based in MOP. It can be expressed in

$$\text{MF(t)} = (\text{CM} + \text{PM})(\text{t})/(\text{p}) \tag{2}$$

where t is a time frame of the period (this research is used "week" as time frame) and p is number of events in the period, in this case 52 week per year.
Example for t = 1,

$$\text{MF}(1) = (121 + 180)/52 = 5.79$$

- Ending Order (EO)(t) is based on information at end of work order in shop, in a period t.
- Starting Inventory (SI) is the quantity of the stock at the end of the period before:

$$\text{SI(t)} = \text{EI(t} - 1) \tag{3}$$

Example for t = 2:

$$\text{SI}(2) = \text{EI}(1) = 0$$

- Ending Inventory (EI) is the quantity of equipment after processing the quantity that arrived and quantity that was used:

$$\text{EI(t)} = \text{SI(t)} + \text{EO(t)} + \text{RO(t)} - \text{MF(t)} \tag{4}$$

Example for t = 3:

$$EI(3) = 0 + 0 + 5.79 - 5.79 = 0$$

- Receiving Order (RO) is when the Maintenance Order will finish and is ready to use. It can be expressed:

$$RO(t) = (MF + SS)(t) - (EO + SI)(t) \tag{5}$$

Example for t(2):

$$RO(2) = (5.79 + 0)(2) - (0 + 0)(2) = 5.79.$$

RO only can be processed if there is a time period available in function of MT. In RO(1) is 0 because it is not possible to process a maintenance in the same period because the MT = 1.

- Work Order (WO) is the moment that the service order is sent to the shop office to do maintenance. This order is:

$$WO(t) = RO(t + MT) \tag{6}$$

Where MT is maintenance time in week. In this example is 1 week.
Example for t = 1:

$$WO(1) = RO(1 + 1) = 5.79$$

- PM Order (PWO) is calculated by multiplying the Work Order and the proportion of preventive maintenance over the total of maintenance in a year. It can be expressed:

$$PWO(t) = WO(t) * PM/(PM + CM)(y) \tag{7}$$

Example for t = 1 and y = y:

$$PWO(1) = 5.79 * (180/(121 + 180)(y)) = 5.79 * 0.6 = 3.47$$

- CM Order (CWO) is calculated by multiplying the Order and the proportion of corrective maintenance over the total of maintenance in a year. It can be expressed:

$$CWO(t) = WO(t) * (CM/(PM + CM))(y)) \tag{8}$$

Example for t = 1:

$$CWO(1) = 5.79 * (121/(121 + 180)y) = 5.79 * 0.4 = 2.33$$

Table 5 Master Maintenance Planning Schedule–MMPS to Reparable

		Year	y − 1	y			
Generator		Period	52	1	2	3	4
Parameters		Maintenance Forecast (MF)		5.79	5.79	5.79	5.79
Maintenance time (MT)	1	Ending Order (EO)		5.79			
Lot Size	1	Starting Inventory (SI)		0	0	0	0
Safety Stock	0	Ending Inventory (EI)	0	0	0	0	0
		Rec. Order (RO)		0	5.79	5.79	5.79
Proportion		Work Order (WO)		5.79	5.79	5.79	
PM	CM	PM Order (PWO)		3.47	3.47	3.47	
0.6	0.4	CM Order (CWO)		2.33	2.33	2.33	

The information of PWO and CWO is transferred to MMRP and CMMS at the end of each period; the system recalculates the quantity again. The sequence of the events in a year or in week time frame 1–4 is in Table 5.

3.4.1 Maintenance Material Requirement Planning–MMRP

After the system generates the Schedule and Corrective planning of Maintenance in MMS, the MMRP function can generate the Material Purchase Planning. In this Example, the Part Number A is used in preventive and corrective maintenance of the generator. In the Preventive Maintenance, the average used is 10, and the corrective maintenance is 7.

The Quantity per Maintenance (QM) is calculated by the average of material that is used in the preventive (QMP) and corrective maintenance (QMC). This information comes from CMMS. Planning Module consolidates the information and sends it to MMRP.

The elements of demand of Part Number "A" of MMS are:

- Preventive Order Demand (POD) represents the material that is used in any preventive maintenance per reparable. It can be expressed:

$$\text{POD(t)} = \text{QMP} * \text{PWO(t)}; \tag{9}$$

Example for t = 1:

$$\text{POD(1)} = 10 * 3.46 = 34.6$$

- Corrective Order Demand (COD) represents the material that is used in any corrective maintenance per reparable. It can expressed:

$$\text{COD(t)} = \text{QMC} * \text{CWO(t)} \tag{10}$$

Table 6 Consolidate Demand of Spare Parts

Part A		Year	y-1	y			
Generator Maintenance	QM	Week Number	52	1	2	3	4
Preventive	10	PO Demand (POD)		34.6	34.6	34.6	34.6
Corrective	7	CO Demand (COD)		16.3	16.3	16.3	16.3
		Total Demand (TOD)		50.9	50.9	50.9	50.9

Example for t = 1:

$$\mathrm{COD}(1) = 7 * 2.33 = 16.29$$

- Total demand (TOD) is the sum of the demand in a time frame:

$$\mathrm{TOD(t)} = \mathrm{POD(t)} + \mathrm{COD(t)} \tag{11}$$

Example for t = 1:

$$\mathrm{POD}(1) = 34.6 + 16.3 = 50.9.$$

All calculations can be seen in the Table 6.

When the demand is consolidate is possible to calculate the material to purchase. In this example the stock starts with 51.4. The calculation can be seen at Table 7 As was discussed, regarding the lot size used in MRP, this research chose to use EOQ because the computational cost is low and the total cost of inventory is near the other models explained by [35].

Table 7 MMRP of Part A

		Year	y-1	y				
Part A		Week Number	52	1	2	3	4	5
		Total Demand (TOD)		50.9	50.9	50.9	50.9	50.9
Lead Time (LT)	4	Ending Requisition (ER)			155			
Lot Size (LS)	155	Starting Inventory (SI)		51.4	0.5	104.6	53.7	2.8
Safety Stock (SS)	0	Ending Inventory (EI)	51.4	0.5	104.6	53.7	2.8	106.9
EOQ	155	Receiving Requisition (RR)		0	0	0	0	155
		Purchasing Requisition (PR)		155				

The following assumption is used to calculate EOQ. The average of demand in a period of 1-year ($\overline{D}$), K is the fixed cost and H is the holding cost. The EOQ formula is:

$$EOQ = \sqrt{\frac{2K\overline{D}}{H}} \tag{12}$$

The safety stock (SS) is service level required (z), multiplies for the standard deviation in a period of 1 year (STD), and square root of the lead time (Lt).

$$SS = Z * STD * \sqrt{Lt} \tag{13}$$

In the example, the item has a fixed cost of \$ 50.00 and the Holding Cost for week is equal the price of the item (\$20.00) multiplied by the annual rate of 22 %. Transforming this rate per week, the holding cost is \$ 0.21 and the Lead-Time is 4 weeks. The average of demand of 1 year is 50.90. So, the result is:

$$\text{EOQ} = \sqrt{\frac{2 * 50 * 50.90}{0.21}} = 154.56$$

SS = 0 because STD is 0 in this example.
Lot size = roundup EOQ = 155
The elements of MMPS are:

- Total Demand (TOD) is the sum of demand at Table 6.
- Ending Requisition (ER) is the information when the requisition is active and when the material will arrive. This information comes from TMS and PMS.
- Starting Inventory (SI) is the quantity of the stock at the end of the period before:

$$\text{SI(t)} = \text{EI(x} - 1) \tag{14}$$

Example for t = 2:

$$\text{SI(2)} = \text{EI(2} - 1) = 0.5$$

- Ending Inventory (EI) is the quantity of material after processing the quantity that arrived and quantity that is used. It can be expressed:

$$\text{EI(t)} = \text{SI(t)} + \text{ER(t)} + \text{RR(t)} - \text{TOD(t)} \tag{15}$$

Example for t = 1:

$$\text{Ex : EI(1)} = 51.4 + 0 + 0 - 50.9 = 0.5$$

- Receiving Requisition (RR) is when the Requisition Order will finish and is ready to use. This time is used to make the decision to order or not.

$$\text{If } SI(t) + ER(t) - TOD(t) < SS(t), \text{ then } RR(t) = EOQ \tag{16}$$

Example t = 5:

$$SI(5) + ER(5) - TOD(5) < SS(5) \geq (2.8 + 0 - 50.9) < 0, \text{ so } RR(5) = 155.$$

This function can only be processed if the lead-time permits.

- Purchasing Requisition (PR) is the moment that the purchase order is sent to the supplier. It can be expressed:

$$PR(t) = RR(x + Lt) \tag{17}$$

Where Lt is lead time. In this example Lt = 4.

Example for t = 1

$$PR(1) = RR(1 + 4) = R(5) = 155$$

The sequence of the events in a year or in week time frame 1–5 is in Table 7

4 Methodology

This section presents the research question with hypotheses and describes the experiment designed to answer the question.

The purpose of this experiment is to test a new collaboration model between maintenance supply chain elements, to match inventory to maintenance requirements and to decrease inventory cost. This research is important because the result tries to reduce uncertainty and consequently, to decrease cost and increase the availability of the equipment.

This investigation applies information processing theoretical approach to analyze the integration of information between the elements of the maintenance supply chain. It expands the idea that with the new technology and techniques (e.g., ERP), that if the new model connects the elements of supply chain, then it can increase the capacity of information processing and consequently decrease uncertainty and costs. The specific research question addressed in this chapter is:

Does Maintenance Enterprise Resource Planning (MERP) decrease inventory costs compared with the EOQ model?

To answer this question, the experiment will test seven hypotheses:

H–1: There is significant difference between different inventory models and quantities of maintenance on inventory cost.

H-2 to H-7 (to each level of maintenance): Inventory cost is lower using MERP than the EOQ model with different quantities of maintenance.

4.1 Independent Variable

- *Inventory Model*: represents the rule that managers can use to decrease the costs associated with maintaining an inventory and meeting customer demand [23]. There are two nominal levels for this variable.

1. Maintenance Enterprise Resource Planning (MERP)-represents a model that increases the capacity to process information by connecting the elements of the supply chain to work as a system. The model was explained in the preceding section.
2. Economic Order Quantity (EOQ)–Harris [20] created a model that seeks to minimize the order cost and holding costs [20]. This is one of earliest and most well-known inventories [31].

EOQ model uses the following formula:

$$EOQ = \sqrt{\frac{2KD}{H}} \tag{18}$$

EOQ = order sizes in units, D = total demand in unit period, H = cost to hold a unit per period of time, K = accounts for when an order is placed [32, p. 33].

In this experiment, the demand will be sum of demand in one year before the period of planning.

This model represent a continuous review policy (Q,R), whenever inventory levels fall to reorder level (ROP) an order for Q units is placed [32]. The ROP has two factors: First is the average of demand ($\bar{D}$) during lead-time (Lt), and second is the safety stock (SS), which is the "amount of inventory that the distributor needs to keep at the warehouse to protect against deviations from the average demand during lead time" [32, p. 42].

$$ROP = Lt * \bar{D} + z * STD * \sqrt{Lt} \tag{19}$$

z is a constant associated service level and STD is standard deviation of average demand in the period.

- *Quantity of Maintenance*: represents a quantity of maintenance that will be performed in a period.

The maintenance can be measured by frequency or elapsed time. This experiment will use the quantity of maintenance by elapsed time (e.g., aircraft maintenance occurs after 100 h flown, Generator TBO occurs after 3,000 h flown).

To change the quantity of maintenance in this experiment, manipulate the quantity of hours per month that an aircraft flies. The range of this variable uses equal interval scales that will vary from very low to high. High represents when an aircraft flies

Table 8 Level of Quantity of Maintenance

Range	Quantity
High-H	205
Medium High-MH	165
Medium-M	125
Low Medium-LM	85
Low-L	45
Very Low-VL	5

internationally; on average it represents 12 h per day. Generally, airplane flies six days a week (48 h), and monthly, (192 h). So the research starts the range (very low) with 5 h monthly, and increases with interval of 40 h until reaching 205 h. The range can be seen in the in the Table 8.

The research will simulate the inventory cost of each model having high or low maintenance. The intention is to check how the models affect the inventory cost with high or low material consumption in an uncertain maintenance environment.

This way, no matter which reparable or material consumption used, the importance is with the range of the amount of maintenance and the behavior that the stock will have. Thus, the experiment is intended to cover the full range of maintenance and material consumption possible and analyze it in each inventory model.

4.2 Dependent Variable

- *Inventory Cost*: the dependent variable is inventory cost. To calculate the inventory cost, this research uses three components: holding cost, fixed cost and shortage cost.

1. Fixed cost: K is accounted, every time that it is placed an order;

$$\mathrm{Ck} = \mathrm{K} * \mathrm{N} \tag{20}$$

N quantity of order in a period.

2. Holding Cost(h): also referred to as a inventory carrying cost, “is accumulated per unit held in inventory per day that the unit is held” [32]. Ballou and Srivastava [1] affirms that 80 % holding costs is referred to as a capital cost [1, p. 348]. Cost of capital can vary from 5 to 35 %. Others variable costs compose the holding cost such as insurance, shelf life limitations and operating cost involved storing inventory or cost of operating warehouse facility [35, p. 138]. In this research will use annual Holding Cost Per Unit:

$$\mathrm{Ch} = \mathrm{C} * \mathrm{H}(\text{in \$/item in inv./year}) \tag{21}$$

Table 9 Total cost calculate

	Sum of qty negative stock in a period	qty ordered in a period	Sum of qty positive stock after in a period
Qty	100	39.00	21,360.10
Parameters	P=21.6	K=54	h=0.4
Total Cost	Shortage Cost	Order Cost	Holding Cost
12,810.04	2,160.00	2,106.00	8,544.04

Table 10 Factorial design of experiment

Independent variable	Inventory models	
Quantity of maintenance	EOQ	MERP
High	Inventory cost	Inventory cost
Medium-high	Inventory cost	Inventory cost
Medium	Inventory cost	Inventory cost
Low-medium	Inventory cost	Inventory cost
Low	Inventory cost	Inventory cost
Very low	Inventory cost	Inventory cost

3. Shortage Cost—occurs when demand exceeds the available inventory for an item. It is related to the level of customer service that the organization wants to reach. It can be like a missed chance of profit, which is called the opportunity cost. In this research, this cost is the quantity missed (S) of item in period times the price of the item (P):

$$\mathrm{Cs} = \mathrm{P} * \mathrm{S} \tag{22}$$

4. Total cost (TC): is the sum of the there components: fixed, holding and shortage cost. It is represented in the following formula:

$$\mathrm{TC} = \mathrm{Ck} + \mathrm{Ch} + \mathrm{Cs} => \mathrm{TC} = \mathrm{K} * \mathrm{N} + \mathrm{H} * \mathrm{Q} + \mathrm{P} * \mathrm{S} \tag{23}$$

An example of the calculation is at Table 9.

The factorial design 2 × 6 of the experiment is represented in Table 10.

4.3 Simulation Experiment

To compare the effect of the models over inventory cost, we do a simulation experiment with empirical data. This empirical experiment controls all internal threats and seeks to study the relations "under a pure and uncontaminated condition" [25, p. 581].

The result of the experiment is compared and analyzed to support the hypotheses, or not. The experiment design is:

Situation A_1^n————X(EOQ)—————O

Situation A_1^n————X(MERP)—————O

n is the number of sample per quadrant in factory design.

Basically the purpose of the simulation experiment is to test the hypotheses derived from the theory. The weakness of generalizing the hypotheses is compensated for per strong internal validity [25]. The simulation represents the reality of an environment. The simulation manipulates the independent variables and records the dependent variable to analyze. This kind of experiment allows for "all of the roles of the research scientist without having to contend with the time-consuming process of data collection" [2].

The time of the experiment is of 4 years, (y −2, y −1, y, y +1). In each year, it will set up the weekly average usage to process the quantity of maintenance. In y −2 and y −1, it will calculate the demand of corrective and preventive maintenance, the spare part consumption of the maintenance, and the weekly average. For the y, and y +1 are simulated 52 events for year with total 104 events for sample. Then, the result experiment is recorded.

The Simulator was programmed using Visual Basic for application along with Microsoft Excel. The Excel is used to produce a useful and comfortable tool [22]. It permits easy testability and repetition of the experiment. The simulation was programmed to produce 50 samples in each quadrant of the factorial design. The simulation ultimately creates 600 samples.

The simulator utilizes a lot of Excel worksheets to process, record and analyze the information. The first step is to fill in the variable and fix parameters. With this information, the quantity of PM and CM per year (MOP function) are calculated. Based on the weekly average of maintenance, the simulator creates a random Poisson number/quantity of maintenance per week to represent the uncertainty.

For an EOQ simulation, the material consumption used in maintenance is processed and calculated for the EOQ (EOQ Demand is the sum of 52 week -1 year-old demand before of actual period week of calculation; ROP uses the average of demand in this period). With EOQ and ROP data, the experiment simulates 2 years of consumption and replacement of stock. To decrease the stock weekly and increase uncertainty, the simulation uses a random Poisson distribution to calculate the consumption of material. In the end, simulator records the EOQ costs.

For MERP, it is uses the same data of maintenance (MOP) and generate a MPS with the quantity of PM and CM. Afterwards, it generates the spare parts to purchase based on MMRP. To decrease the weekly stock and increase uncertainty, the simulation uses a random Poisson distribution to calculate consumption of material.

At the end of each procedure, the EOQ and MERP cost and quantities are recorded and the simulator repeats the experiment fifty times with random maintenance and consumption of material. After recording 50 samples, the simulator changes the parameters and processes again until finishing the last parameter. The procedure is in Fig. 3.

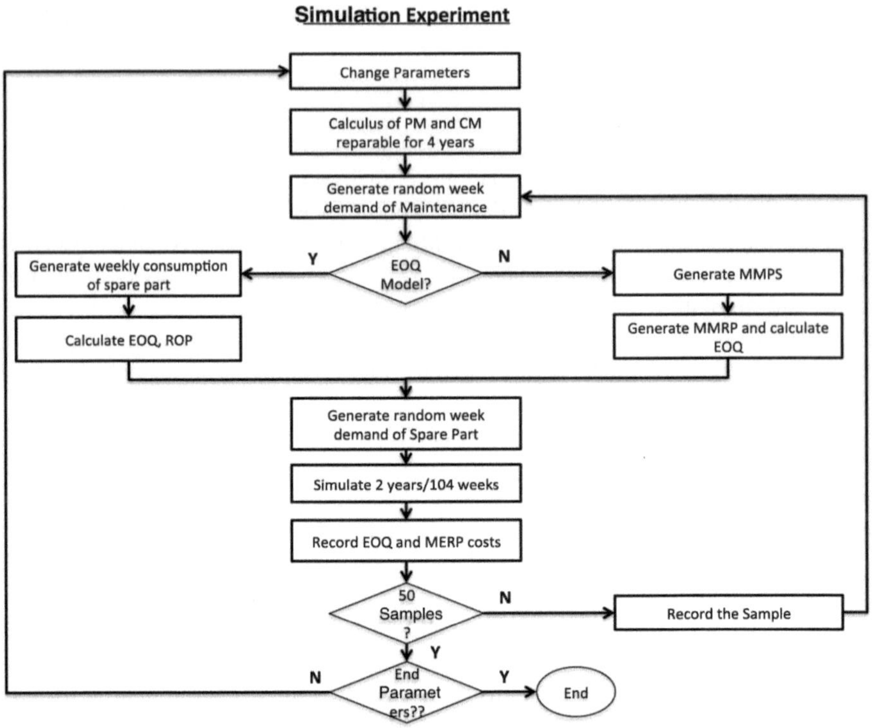

Fig. 3 Simulation procedure

For the H-1, the samples are statically tested with an analysis of the variance (ANOVA) to support the hypothesis that there is a significant difference between the two models. For H-2 to H-7, because the samples are paired, (i.e., the simulation uses the same parameters to produce results in EOQ and MERP), it will use a t-test to check the hypotheses.

4.4 Assumption and Fixed Parameters of the Simulation

- Assumptions for MERP:

  1. The simulation calculates the requirements at the beginning of period.
  2. The simulation tries to meet requirements for future periods;
  3. The decisions occur weekly.
  4. The cost does not change significantly with time.
  5. Supplier delivers the requirement on time; deliveries don't have uncertainty.
  6. The experiment puts uncertainty only on demand requirement (requirement for more or less than planned using Poisson distribution)

Table 11 Fixed parameters

Parameter	Value
Fixed cost (K)	$54.00
Item price	$20.00
Tax annual holding cost(H)	$22 %a.a
Number of aircraft X	300
QPA of generator in aircraft X	2
QPA of Part A in Preventive maintenance of Generator	QPA = 10 Probability of change = 100 %
QPA of Part A in Corrective maintenance = 10	QPA = 10 Probability of change = 80 %
Service level	0.90
Lead time spare part	4 weeks
Frame time of experiment	52 week /year

- Assumption for EOQ:

  1. It uses a continuous review policy for purchases.
  2. ROP and EOQ use historic demand for 1 year.

  Fixed parameters are at Table 11.

5 Results

- Hypothesis 1 - There is significant difference between MERP and EOQ inventory models, and different quantities of maintenance on inventory costs.

To test this experiment, the simulation generates 50 results to each quadrant of a factorial design. Analysis of variance (ANOVA) was used to compare the systematics variance in the data to the amount of unsystematic variance and presents the result at Table 12.

ANOVA produces a F-statistic or F-ratio to support that the means of the experiments are equal or not. The significance level tested is 95 %. The test is at Table 13.

After analyzing the results, researchers can infer that:

- There is a significant main effect of the type of inventory model on inventory cost, $F(5,588) = 470.26$, $p < 0.001$, $\omega^2 = 0.78$.
- There is a significant main effect of the quantity of maintenance on inventory cost, $F(1,588) = 579.94$, $p < 0.001$, $\omega^2 = 0.19$.
- There is a significant interaction effect between inventory models and quantity of maintenance on the inventory cost, $F(5, 588) = 14.30, p < 0.001, \omega^2 = 0.02$.

Table 12 Result of experiment after simulation

Independent variable	Inventory models	
Quantity of maintenance	EOQ	MERP
High	5,604.73	3,845.59
Medium-high	5,250.52	3,451.92
Medium	4,371.84	2,996.56
Low-medium	3,711.36	2,489.34
Low	2,686.83	1,833.45
Very low	1,265.09	685.99

Table 13 ANOVA table

Source	SS	df	MS	F	p-value
Models	972,559,674	5	194,511,934	470.26	9.32E-203
Qty Maintenance	239,876,513	1	239,876,513	579.94	1.11E-89
Interaction	29,570,294	5	5,914,058	14.30	3.18E-13
Error	243,210,640	588	413,623		
Total	1,485,217,122	599			

This indicates that EOQ and MERP models are affected differently by quantity of maintenance.

ω^2 represents the variance estimate for the effect divided by the total variance [13, p. 446].

The result supports Hypothesis 1 that there is significant difference between the two inventory models, and quantity of maintenance on inventory cost.

- H-2 to H-7 (to each level of maintenance): Inventory cost is lower using MERP than the EOQ model with different quantity of maintenance.

After Simulator produced 50 samples to each level of maintenance, it was done a dependent t-test to $p \leq 5\,\%$. With result at Table 14, researchers can infer that on average the experiment present that the inventory cost is significant lower *using MERP than the EOQ model with different quantity of maintenance* according Table 14.

Effect size (r) is " simply an objective and (usually) standardized measure of the magnitude of observed effect" [13, p. 56]. The formula to calculate the effect size is:

$$(2)\, r = \sqrt{\frac{t^2}{t^2 + df}} \tag{24}$$

6 Discussion

The study tests a new collaboration model between maintenance supply chain elements. It matches inventory to maintenance requirements in order to decrease inventory costs. We compare the new model with the traditional inventory model and at

Table 14 H-2-H-7-Dependent t-test

Maintenance Qtt	Parameters	EOQ	MERP
High	Mean	5,604.73	3,845.59
	Std. Error Mean	150.51	18.89
	mean difference (MERP - EOQ)	1,759.14	
	Std. dev.	1,091.74	
	Std. error	154.40	
	t-test	11.39	
	p-value (one-tailed, lower)	1.11E-15	
	r (effect size)	0.85	
	Confidence interval 95. % lower	1,448.87	
	Confidence interval 95. % upper	2,069.41	
	Margin of error	310.27	
Medium-High	Mean	5,250.52	3,451.92
	Std. Error Mean	178.77	16.71
	Mean difference (MERP - EOQ)	1,798.60	
	Std. dev.	1,283.77	
	Std. error	181.55	
	t-test	9.91	
	p-value (one-tailed, lower)	1.37E-13	
	r (effect size)	0.82	
	Confidence interval 95. % lower	1,433.76	
	Confidence interval 95. % upper	2,163.45	
	Margin of error	364.84	
Medium	Mean	4,371.84	2,996.56
	Std. Error Mean	135.19	14.03
	Mean difference (MERP - EOQ)	1,375.27	
	Std. dev.	967.83	
	Std. error	136.87	
	t-test	10.05	
	p-value (one-tailed, lower)	8.59E-14	
	r (effect size)	0.82	
	Confidence interval 95. % lower	1,100.22	
	Confidence interval 95. % upper	1,650.33	
	Margin of error	275.06	
Low-Medium	Mean	3,711.36	2,489.34
	Std. Error Mean	112.47	10.79
	Mean difference (MERP - EOQ)	1,222.02	
	Std. dev.	800.92	
	Std. error	113.27	
	t-test	10.79	
	p-value (one-tailed, lower)	7.62E-15	
	r (effect size)	0.84	

(continued)

Table 14 (continued)

Maintenance Qtt	Parameters	EOQ	MERP
	Confidence interval 95. % lower	994.40	
	Confidence interval 95. % upper	1,449.64	
	Margin of error	227.62	
Low	Mean	2,686.83	1,833.45
	Std. Error Mean	79.35	8.42
	Mean difference (MERP - EOQ)	853.38	
	Std. dev.	559.40	
	Std. error	79.11	
	t-test	10.79	
	p-value (one-tailed, lower)	7.67E-15	
	r (effect size)	0.84	
	Confidence interval 95. % lower	694.40	
	Confidence interval 95. % upper	1,012.36	
	Margin of error	158.98	
Very Low	Mean	1,265.09	685.99
	Std. Error Mean	79.82	6.69
	Mean difference (MERP - EOQ)	579.10	
	Std. dev.	562.68	
	Std. error	79.57	
	t-test	7.28	
	p-value (one-tailed, lower)	1.23E-09	
	r (effect size)	0.72	
	Confidence interval 95. % lower	419.18	
	Confidence interval 95. % upper	739.01	
	Margin of error	159.91	

different quantities of maintenance. The research question is supported by the result of seven hypotheses.

The first hypothesis shows that there are strong differences in inventory costs using models with different quantities of maintenance. Models, quantities of maintenance and their interactions have a significant effect on inventory cost. Although the experiment demonstrates that both models purchase almost the same quantity of material, inventory cost is different between the models when different quantities of maintenance are applied.

The second through the seventh hypotheses are supported by the dependent statistic t-test. The t-test supports that when MERP is used to manage inventory, the cost is lower than with EOQ. Therefore, we can infer that there is strong evidence that Maintenance Enterprise Resource Planning (MERP) model decreases inventory costs when compared to the EOQ model.

This research extends the use of information processing theory to supply chain management by creating a model that integrates information within and across the supply chain. Because of the complexity of the maintenance environment, the model organizes, shares and integrates information among the elements of the supply chain (e.g., MTBUR, BOM, hours of flight). MERP framework increases the integration capability, and consequently, can increase supply chain performance. So, the model extends the Galbraich (1973) proposal where with high uncertainty, there is more need for processing information. This model increases the lateral and vertical integration providing a great increase in cost performance in supply chains. Posey and Bari [29] propose a framework to supply chain but didn't test the framework. This experiment complements the study of [29] by showing results that are proposed in their framework.

This research adds a new scientific approach to MRP by adding a new theory on the use of MRP. In the early days, "MRP was neglected in academic curricula in favor of intellectually challenging statistical and mathematical techniques. Academics considered the study of MRP vocational rather than scientific" [30, p. 375]. This experiment uses the principle of information-processing theory to integrate lateral relation and increase vertical information to decision makers, a principle of MRP. Using MRP techniques, this model can increase the capacity of information processing and decrease uncertainty in the maintenance supply chain.

Further, this model brings a new framework to the maintenance supply chain. A literature review shows scarce research about models that attend to this environment. This model brings a new management dimension to maintenance supply chain. With it, MRO organizations can integrate the use of equipment, predict maintenance and material, and consequently, decrease inventory costs. This framework fits well in organizations that specialize in management maintenance and service supply chain.

Reducing inventory costs can now be explained. The integration of information decreases the degree of separate information, so that there is both a reduction in uncertainty and an increased information processing capacity. "Traditional inventory management, in the pre-computer days, could not process and integrate the information because of limitations imposed by the information-processing tools". Almost all those approaches suffered from this imperfection causing development of elaborate mathematics models working in isolation, such as with the EOQ and ROP models [30, pp. 377–378].

The new model decreases the volume of uncertainty by putting the maintenance demand as a mitigating factor. So, demand forecasting mitigates uncertainty and consequently the quantity of the stock needed to attend the maintenance is lower than the buffer class in EOQ.

This simulation controls the unbiased variables and manipulates the independent variables to measure the dependent variables. This model studies only an aircraft, a generator and a spare part, but the pattern observed in this experiment can be applied to any reparables or spare parts. Only the basic parameters change, yet the results are the same because the models tested the high and low quantities of maintenance demand. So the spare parts have to follow the same pattern for any reparable. This model can be used for all items of an aircraft, and results will be the same. By putting all reparables and spare parts in MERP models, managers can simulate the fleet usage and can adjust the quantity to fit their budget. Nowadays, the only limitations are the processing capacity, which, is easily overcome with the improved capacity of new computers and networks.

This model can also bring new approaches to manage maintenance. For example, car dealers have to maintain a high inventory to attend to corrective and preventive maintenance. If cars now have technologies such as machine-to-machine (M2M) that transmit mileage, MERP can calculate and forecast maintenance and material requirements and decrease the materials inventory for shop maintenance. All companies doing maintenance can use this framework to improve their supply chain.

This research uses uncertainty in demand only. For future research, it is suggested to put uncertainty into lead time, and to study new buffers against such uncertainty such as Demand-Driven MRP [30]. Other useful research would include testing this model in a real environment to record the data and compare it across the simulations performed.

References

1. Ballou RH, Srivastava SK (2007) Business logistics/supply chain management. Pearson, Noida
2. Benedict JO, Butts BD (1981) Computer simulation of real experimentation: is one better for teaching experimental design? Teach Psychol 8(1):35–38. doi:10.1207/s15328023top0801_10
3. Blanchard BS, Verma DC, Peterson EL (1995) Maintainability: a key to effective serviceability and maintenance management. Wiley-Interscience, New York
4. Bojanowski RS (1984) Improving factory performance with service requirements planning (SRP). Prod Inventory Manag 25(2):31
5. Bolon DS (1998) Information processing theory: implications for health care organizations. Int J Technol Manag 15(3,4,5):211–221
6. Cohen MA, Agrawal N, Agrawal V (2006) Winning in the aftermarket. Harv Bus Rev 84(5):129–138
7. DePuy GW, Usher JS, Walker RL, Taylor GD (2007) Production planning for remanufactured products. Prod Plan Control 18(7):573
8. DoD (2009) Supply system inventory report. Office of the under secretary of defense for acquisition, technology, and logistics. Available from http://www.acq.osd.mil/log/sci/exec_info/ssir_new/FY2009_SSIR_MAC_final_update.pdf
9. Ettkin LP, Jahnig DG (1986) Adapting MRP II for maintenance resource management can provide a strategic advantage. Ind Eng 18(8):50
10. Fabry C, Schmitz-Urban A (2010) Maintenance supply chain optimisation within an IT-platform: network service science and management. In: 2010 international conference on management and service science (MASS), pp 1–4. doi:10.1109/ICMSS.2010.5576323
11. Ferrer G, Whybark DC (2001) Communicating product recovery activities. In: Madu C (ed) Handbook of environmentally conscious manufacturing SE-4. Springer, pp 81–99. doi:10.1007/978-1-4615-1727-6_4
12. Ferrer G, Whybark DC (2001) Material planning for a remanufacturing facility. Prod Oper Manag 10(2):112–124
13. Field A (2009) Field, discovering statistics using SPSS, 3e “and” SPSS CD version 17.0. Sage Publications Inc

14. Flynn BB, Flynn EJ (1999) Information-processing alternatives for coping with manufacturing environment complexity. Decis Sci 30(4):1021–1052
15. Gaither N (1983) An improved lot-sizing model for MRP systems. Prod Inventory Manag 24(3):10
16. Galbraith J (1974) Organization design: an information processing view. Interfaces 4(3):28–36
17. Galbraith J (1977) Organization design. Addison Wesley Publishing Company, Reading
18. Gaudette KJ (2003) Inventory planning for remanufacturing. ProQuest Dissertations and Theses. Indiana University, Ann Arbor
19. Ghobbar AA, Friend CH (2007) Aircraft maintenance and inventory control using the reorder point system. Int J Prod Res 34(10):2863–2878
20. Harris FW (1913) How many parts to make at once. Oper Res - Mag Manag 38(6):947–950. doi:10.1287/opre.38.6.947
21. Heizer J, Render B (2007) Principles of operations management, 7th edn. Pearson/Prentice Hall, Upper Saddle River, New Jersey, p 684
22. Hihn J, Lewicki S, Wilkinson B (2009) How spreadsheets get us to mars and beyond. In: 2009 42nd Hawaii international conference on system sciences, pp 1–9. doi:10.1109/HICSS.2009.239
23. Hillier FS, Lieberman GJ (1980) Introduction to operations research, 3rd edn. Holden-Day Inc, San Francisco
24. Jones J (2006) Integrated logistics support handbook. McGraw-Hill Professional
25. Kerlinger FN, Lee HB (1999) Foundations of behavioral research, 4th edn. Wadsworth, New York, p 890
26. Markus ML, Axline S, Petrie D, Tanis C (2000) Learning from adopters' experiences with ERP: problems encountered and success achieved. J Inf Technol 15(4):245–265. doi:10.1080/02683960010008944
27. Molinder A (1997) Joint optimization of lot-sizes, safety stocks and safety lead times in an MRP system. Int J Prod Res 35(4):983–994. doi:10.1080/002075497195498
28. Newman RG (1985) MRP where M = Preventative maintenance. Prod Inventory Manag J 26(2): 21 to eoa
29. Posey C, Bari A (2009) Information sharing and supply chain performance: understanding complexity, compatibility, and processing. Int J Inf Syst Supply Chain Manag 2(3): 67–76. doi:10.4018/jisscm.2009070105
30. Ptak C, Smith C (2011) Orlicky's material requirements planning, 3rd edn. McGraw-Hill Professional, NY, p 546
31. Silver EA, Pyke DF, Peterson R (1998) Inventory management and production planning and scheduling, 3rd edn. Wiley, Hoboken-Nj, p 784
32. Simchi-Levi D, Kaminsky P, Simchi-Levi E (2007) Design and managing the supply chain: concepts, strategies and case studies, 3rd edn. McGraw-Hill Irwin, New York, p 498
33. Swanson L (2003) An information-processing model of maintenance management. Int J Prod Econ 83(1):45–64
34. United States CB (2007) Industry statistics sampler. Available from http://www.census.gov/econ/industry/products/p811.htm, http://www.census.gov/econ/industry/products/p336411.htm
35. Vollmann T, Berry W, Whybark DC, Jacobs FR (2005) Manufacturing planning and control for supply chain management, 5th edn. McGraw-Hill/Irwin, New York, p 709
36. Wemmerlov U, Whybark DC (1984) Lot-sizing under uncertainty in a rolling schedule environment. Int J Prod Res 22(3):467
37. Whybark DC, Williams JG (1976) Material requirements planning under uncertanty. Decis Sci 7(4):595

Part IV
Courier, Express, and Parcel Service Network Design

Strategic Planning of Optimal Networks for Parcel and Letter Mail

Martin Nikolas Baumung, Halil Ibrahim Gündüz, Thomas Müller and Hans-Jürgen Sebastian

Abstract This paper considers postal logistics, more precisely, the distribution networks for letter mail and parcel mail. The main service provided by postal companies is letter mail and parcel mail transportation and delivery. In this market segment there have been two key efforts during the last few years: reduction in transportation and delivery time (service quality) and minimization of costs. Both efforts—reduction of service time and minimization of costs for providing the promised services—have a strong impact on the quality of the strategic planning phases of the respective distribution networks. In this article we introduce the structure of a typical distribution network for letter mail and for parcel mail, and we describe the main subnetworks. Furthermore, this paper deals with two selected projects on optimization of such networks. Each of the projects covers system analysis, modeling, and for the second project, also the development of an optimization algorithm.

1 Introduction

The increasing market competition and the service focus of customers force logistics service providers, such as postal organization and express shipment companies, to re-evaluate and to continuously improve their networks for parcel, letter, and freight mail. The core service provided by postal companies is parcel and letter mail transportation and delivery.

M.N. Baumung (✉) · H.I. Gündüz · T. Müller · H.-J. Sebastian
Deutsche Post Chair of Optimization of Distribution Networks, RWTH Aachen University, Kackertstr. 7, 52072 Aachen, Germany
e-mail: baumung@dpor.rwth-aachen.de

H.I. Gündüz
e-mail: guenduez@dpor.rwth-aachen.de

T. Müller
e-mail: mueller@dpor.rwth-aachen.de

H-J. Sebastian
e-mail: sebastian@dpor.rwth-aachen.de

H.-J. Sebastian et al. (eds.), *Quantitative Approaches in Logistics and Supply Chain Management*, Lecture Notes in Logistics,
DOI 10.1007/978-3-319-12856-6_4

Worldwide, the parcel mail market is rapidly growing, while the volume of letter mail is decreasing. This situation causes changes in the distribution networks.

1.1 Postal Network Design

A typical distribution network for parcel or letter mail can be described briefly by the following subnetworks (see Fig. 1):

Mail collection subnetwork: In this network, mail (i.e. parcels or letters) is collected from different mail sources (e.g. mailboxes, business customers) and transported to sorting centers. Consolidation points (CoP) are used to switch from small vehicles to bigger vehicles, which then transport the mail to the sorting centers.

Sorting centers (SC): Sorting centers are big automated sorting facilities for parcel or letter mail, which work in two different modes during different time intervals. The input sorting center (ISC) performs sorting with respect to the destination sorting center (SC) (in Germany, characterized by the first two digits of the 5-digit zip-code), while the output sorting center (OSC) performs sorting processes for the distribution and delivery to the final destination.

Long-haul transportation subnetwork: The subnetwork takes care of the mail exchange between the sorting centers (overnight). The main idea is to use bigger and faster vehicles for long distances and to consolidate mail at a subset of sorting centers which are used as hubs. In real world applications, the long-haul transportation subnetwork is often realized in two different transportation modes (e.g. air–road, road–rail).

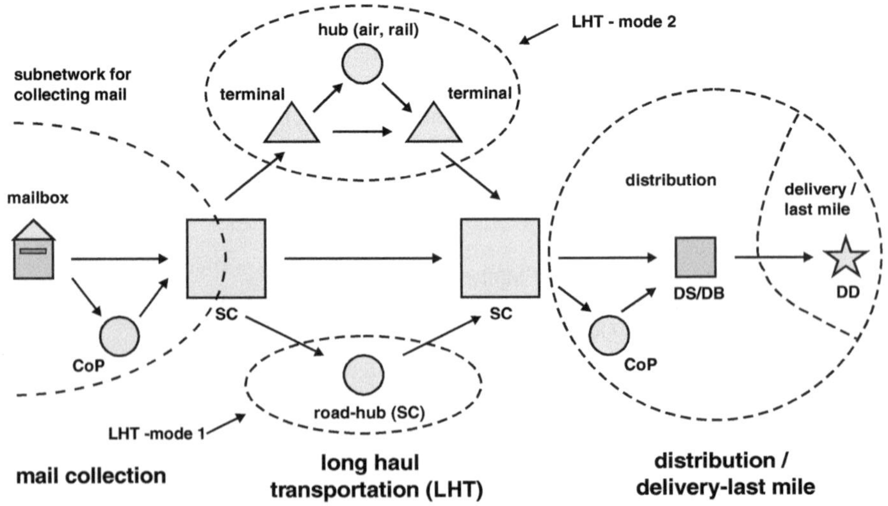

Fig. 1 Components/Subnetworks of a distribution network for letter and parcel mail

Distribution subnetwork: In this subnetwork, the mail is distributed by transportation from the sorting centers to mini-hubs (called: delivery stations (DS) in the case of letter mail, and delivery bases (DB) in the case of parcel mail) using vehicle routes. At the mini-hubs a final sorting in sequence takes place for each of the assigned delivery districts (DD).

Delivery (last mile) subnetwork: Each postman has an allocated delivery district. He/She starts the delivery route at the mini-hub, moves to the delivery district and visits it in a predefined (optimal) sequence and finally returns to the mini-hub (DS/DB).

This brief description of the complex networks in postal logistics suggests that the problem of optimizing such a network may become complicated. Of course the instances of such networks in different countries differ significantly. For example, an instance of the DHL's parcel mail network in Germany in 2011 is characterized by:

- 33 sorting centers (parcel mail centers), 203 delivery bases, 790 million parcels per year, 7,600 vehicles for the parcel mail delivery, 7,500 delivery districts.
- In 2011, the letter mail network of Deutsche Post in Germany consisted of: 82 sorting centers (plus the international sorting center), 3,100 delivery stations, 66 million letter mails per day, 53,000 delivery districts, 80,000 postmen, 110,000 mail boxes.

Although the networks for parcel and letter mail are separated, there are synergies. The delivery of parcels and letters is realized together (same postman, same vehicle) for a large number of delivery districts. Further, the distribution of parcels and letters is done by the same vehicles if this makes sense.

From this data it follows that the optimization problem is very complex for both German networks of DPDHL. In addition, the acquisition of all data which is needed to feed the respective optimization models requires methods from data and system analysis.

For such complex problems, it is well known that introducing planning phases is necessary in order to reduce complexity and to deal with a collection of models with different time and resource granularity.

In this paper we mainly focus on the strategic planning phase. The strategic planning phase deals with long-term decisions related to the network infrastructure:

- decisions related to the quantity and quality of the main resources (locations, facilities, vehicles, human resources) and the method of acquisition of these resources and
- the selection of services to be offered.

1.2 Problems Studied

We will consider problems and models which originate in the postal logistics area (outlined briefly above), on a more generic level for applications in different areas.

Further, we will discuss one real-world application from DPDHL for each generic model.

The first group of generic models deals with the strategic optimization of parcel mail distribution networks. The real-world application is DHL's national parcel mail network for Germany. The key to the success of this approach is the modeling of a sequence of three optimization models, starting with the simplest case, and elaborating models which are becoming more and more complicated and detailed. In the following these three models are briefly described, however only the first-stage model is covered in this paper while the two other models will be subjects of future works.

The first-stage model is the location/allocation model, which selects sorting centers from a given finite set of candidates and allocates geographic areas to the selected sorting centers, where the allocated areas are composed of 5-digit zip-code areas. The objective is to minimize costs. The solution represents the overall design of the parcel mail network. The service quality, which can be reached with this cost-minimized network, is computed after the strategic optimization by approximating the transportation process within this network. The model was implemented using the modeling language AIMMS [1] and the DHL problem instances have been solved to optimality by CPLEX [11].

The second-stage model adds service quality constraints (e.g. next day delivery of 90 percent of all products) to the cost minimization model. Unfortunately, this requires time as an explicit factor in the optimization model in order to estimate transportation time from the network sources to the final destination. This model cannot be solved to optimality using solvers, available today. Therefore, we developed metaheuristics to solve the optimization problems.

The third-stage models are the computationally most challenging models in this collection. Optimization of the locations of delivery bases (i.e. the mini-hubs of a parcel mail network) requires that the sorting centers and their allocated areas are given. The approximate solution of this third-stage model for DHL's problem instance leads to an interesting application with a high potential for cost savings.

The second group of models relate to a long-haul transportation subnetwork with hubs. In the parcel mail network discussed above, no hubs are used between the parcel mail centers within the long-haul transportation network. This is the main reason why the overall strategic optimization problem is tractable. However, the long-haul transportation subnetwork between the sorting centers is much more complicated in letter mail distribution networks, such as the Deutsche Post's national letter mail transportation network, because consolidation with hubs is necessary. It shows that the well-known hub location models cannot be used to adequately model this problem.

The solution of the hub location problem is the key to the strategic optimization of the overall network in the case when subsets of sorting centers are used as hubs in the long-haul transportation subnetworks.

We have developed a new model, which is "trip-based" rather than "flow-based", and a collection of heuristics and metaheuristics. In this paper we will show that this approach is able to solve the hub location problem for the long-haul transportation network for letter mail within Germany.

1.3 Literature Review

In this paper we mainly focus on the strategic network design of mail service providers. In the context of optimization, network design problems usually consist of two interrelated problems: to determine the number and locations of hubs and depots and to allocate geographic areas to these facilities. These problems either belong to the class of hub location or to the group of p-hub median problems. Hub location models are well studied in literature and can be found in [7, 26] or [29]. A literature review and some recent trends are given in [2]. Efficient algorithms for the single allocation p-hub median problem can be found in [16]. Further discussion of the related p-hub center problem is introduced by [15, 24].

Network design problems have a lot of real world applications, i.e. railroad or airline network design. On a generic level, [37] studied the optimization of logistic networks considering strategic, tactical and operational costs and developed a corresponding tabu search algorithm. Applications of network design problems to postal logistics have been studied in several different countries [4] restructured the logistic system of the Swiss parcel delivery network and decided upon the number, location, capacity, and service areas of different transshipment points [36] present a heuristic solution concept for the design of the hub transportation network for parcel service providers in Austria and [31] optimized highway transportation at the United States Postal Service.

Reference [34] gives an overview on optimization approaches in the strategic and tactical planning of postal networks and [35] gives an insight to the current state of the art applications at Deutsche Post DHL. Operative aspects like the determination of transport routes, etc. for a private parcel service provider in Poland are considered by [28]. Reference [21] studied similar location-routing problems for fast-delivery subnetworks in urban areas. Reference [6] proposed a hybrid tabu search/branch-and-bound algorithm for the direct flight network design problem and [23] developed a Dantzig-Wolfe decomposition approach for optimizing the letter mail flight network of Deutsche Post DHL. Reference [3] created a system to optimize the design of service networks for delivering express packages for UPS. A survey of planning models for long-haul operations was published by [20]. On a more generic level, [38] developed a multi-stage facility location problem with staircase costs, which finds many applications in postal logistics.

The sheer size of Deutsche Post DHL's parcel mail network made it necessary to analyze the network as well as the data. As a result of this analysis, we were able to make meaningful assumptions in order to adapt and simplify the model to make it computationally tractable. As for the letter mail network, the peculiarities of Deutsche

Post DHL's long-haul transportation network inside the letter mail network made it necessary to develop a new approach to economies of scale, which is introduced in Sect. 3.

The remainder of this paper is structured as follows. In Sect. 2 we present a model to optimize the German parcel mail network of DHL along with corresponding results. In Sect. 3 we focus on optimizing the long-haul transports in Deutsche Post DHL's letter mail network. We propose a generic model which considers the peculiarities of this subnetwork, and we describe a metaheuristic solution approach and provide some results. The paper ends with a conclusion in Sect. 4.

2 Modeling and Optimizing the German Parcel Mail Network of DHL

A comprehensive and nationwide cost analysis of parcel mail distribution was last performed in the years 1995–1997 after the European liberalization of the telecom market. During that time, the whole distribution network of Deutsche Post DHL was restructured. Since this reorganization, the number and location of sorting centers have slightly changed. Hence, a complete analysis of the current and the future parcel mail distribution network is necessary due to the rapidly increasing B2C market. Between 2000 and 2011, the parcel mail distribution in Germany recorded an expansion of 50 % (see Fig. 2).

Since 2007, revenue in the parcel business has increased on average by 3.5 % annually and it went up by as much as nearly 10 % in the first half of 2011. As a result, this growing market already makes up around one-fifth of the total revenue [14]. The analysis of the parcel distribution network is therefore an important step to ensure that the parcel business continues to contribute earnings. The main focus of this section is to find the optimal number and location of sorting centers as well as the optimal allocation of geographic areas to them. For this purpose, a first strategic model is described in the next section.

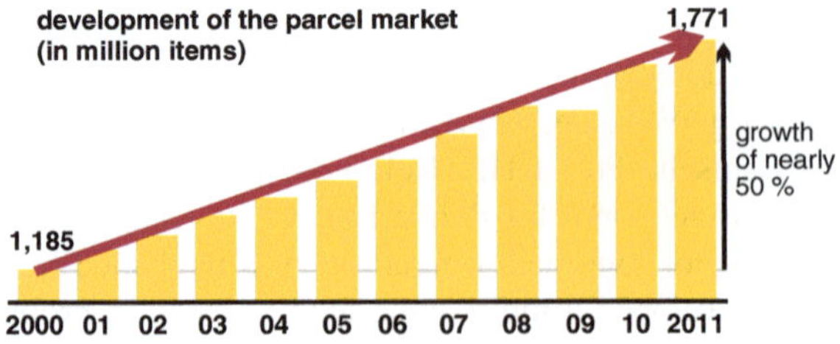

Fig. 2 Development of the parcel mail market from 2000 to 2011 in Germany

2.1 Strategic Model Formulation

The location-allocation model is defined on a weighted (not necessarily complete) directed graph $G = (V, A, C)$. The node set V consists of disjoint subsets S, I, J, and T. S is the set of geographic areas (e.g. 5-digit zip-code areas). Sorting centers are represented by both node sets I and J. While I is the representative set for the potential input sorting centers, J represents the potential set of output sorting centers. Each input/output SC has a sorting capacity K_i^1/K_j^2 and opening fixed costs F_i^1/F_j^2. T is the set of the delivery bases/stations. Further, to each DB/DS, represented by the node set T, the overall parcel mail volume b_t and parcel mail volume b_t^s from each s area is known. The graph contains the following set of arcs (see Fig. 3):

- (s, i) with $s \in S$, $i \in I$ (parcel mail collection)
- (i, j) with $i \in I, j \in J$ (long-haul transportation)
- (i, t) with $i \in I$, $t \in T$ (direct transportation from ISC to DB/DS)
- (j, t) with $j \in J$, $t \in T$ (parcel mail distribution)

C is a function which associates transportation costs to every arc in the set A, where c_{si}^1 denotes the unit costs of transportation of parcel mail collection, c_{ij}^2 the unit costs of transportation of long-haul transportation, c_{it}^3 the unit costs of transportation of direct shipment from ISC to DB/DS, and c_{jt}^4 the unit costs of transportation of parcel mail distribution. The task is to determine the location of open sorting centers, the assignment of areas to open input sorting centers, the assignment of delivery bases/stations to output sorting centers, and the transport flows of the network with minimum overall costs, such that the following constraints hold:

- Each area is assigned exactly to one open ISC.
- Each DB/DS is assigned at most to one open OSC.
- Each collected item of parcel mail must be transported to its destination DB through either long-haul or direct transportation.

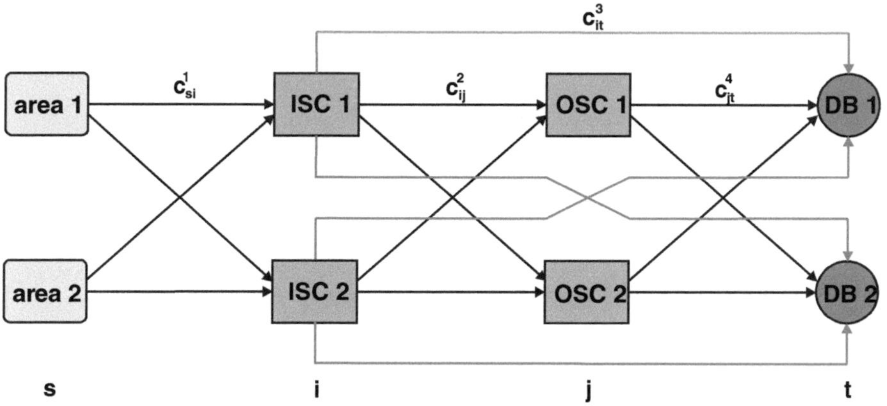

Fig. 3 Schematic representation of the parcel distribution network

- Each incoming transport flow equals the outgoing transport flow of each ISC/OSC.
- The total parcel mail collection of areas assigned to an open ISC does not exceed the sorting capacity.
- The total parcel mail distribution through long-haul transportation to delivery bases/stations assigned to an open OSC does not exceed the sorting capacity.

We introduce the following binary decision variables:

$$x_{si} = \begin{cases} 1, & \text{if area } s \in S \text{ is assigned to ISC } i \in I \\ 0, & \text{otherwise} \end{cases}$$

$$y_{jt} = \begin{cases} 1, & \text{if DB/DS } t \in T \text{ is assigned to OSC } j \in J \\ 0, & \text{otherwise} \end{cases}$$

$$z_{it} = \begin{cases} 1, & \text{if a direct transport from ISC } i \in I \text{ to DB/DS } t \in T \\ & \text{is performed} \\ 0, & \text{otherwise} \end{cases}$$

$$u_i = \begin{cases} 1, & \text{if ISC } i \in I \text{ is open} \\ 0, & \text{otherwise} \end{cases}$$

$$v_j = \begin{cases} 1, & \text{if OSC } j \in J \text{ is open} \\ 0, & \text{otherwise} \end{cases}$$

Further, we introduce the following flow decision variables:

$f_{ij}^t \geq 0$: flow from ISC $i \in I$ to OSC $j \in J$ with destination DB/DS $t \in T$
$f_{it}^3 \geq 0$: flow from ISC $i \in I$ to DB/DS $t \in T$
$f_{jt}^4 \geq 0$: flow from OSC $j \in J$ to DB/DS $t \in T$

A mixed integer linear program can now be stated as follows:

$$\min z = \sum_{s \in S,\, i \in I,\, t \in T} c_{si}^1 \cdot b_t^s \cdot x_{si} + \sum_{i \in I,\, j \in J,\, t \in T} c_{ij}^2 \cdot f_{ij}^t + \sum_{i \in I,\, t \in T} c_{it}^3 \cdot f_{it}^3 \tag{1}$$
$$+ \sum_{j \in J,\, t \in T} c_{jt}^4 \cdot f_{jt}^4 + \sum_{i \in I} F_i^1 \cdot u_i + \sum_{j \in J} F_j^2 \cdot v_j$$

subject to

$$\sum_{i \in I} x_{si} = 1 \quad \forall\, s \in S \tag{2}$$

$$\sum_{j \in J} y_{jt} \leq 1 \quad \forall\, t \in T \tag{3}$$

$$\sum_{s \in S} b_s^t \cdot x_{si} = \sum_{j \in J} f_{ij}^t + f_{it}^3 \quad \forall\, i \in I,\ t \in T \tag{4}$$

$$\sum_{i \in I} f_{ij}^t = f_{jt}^4 \quad \forall\, j \in J,\ t \in T \tag{5}$$

$$\sum_{i \in I} f_{it}^3 + \sum_{j \in J} f_{jt}^4 = b_t \quad \forall t \in T \tag{6}$$

$$f_{jt}^4 \leq b_t \cdot y_{jt} \quad \forall j \in J,\ t \in T \tag{7}$$

$$N \cdot z_{it} \leq f_{it}^3 \quad \forall i \in I,\ t \in T \tag{8}$$

$$M \cdot z_{it} \geq f_{it}^3 \quad \forall i \in I,\ t \in T \tag{9}$$

$$\sum_{s \in S,\ t \in T} b_s^t \cdot x_{si} \leq K_i^1 \cdot u_i \quad \forall i \in I \tag{10}$$

$$\sum_{t \in T} f_{jt}^4 \leq K_j^2 \cdot v_j \quad \forall j \in J \tag{11}$$

$$Min_{\text{ISC}} \leq \sum_{i \in I} u_i \leq Max_{\text{ISC}} \tag{12}$$

$$Min_{\text{OSC}} \leq \sum_{j \in J} v_j \leq Max_{\text{OSC}} \tag{13}$$

$$x_{si},\ y_{jt},\ z_{it},\ u_i,\ v_j \in \{0, 1\} \quad \forall s \in S,\ i \in I,\ t \in T \tag{14}$$

$$f_{ij}^t,\ f_{it}^3,\ f_{jt}^4 \geq 0 \quad \forall i \in I,\ j \in J,\ t \in T \tag{15}$$

The objective function minimizes the sum of parcel mail collection, long-haul, and parcel mail distribution transportation costs, and the sum of fixed costs for opening input/output sorting centers. Constraints (2) guarantee the single assignment of an area to an ISC. Each DB/DS is assigned to at most one OSC through constraints (3). In theory, it is possible that a DB/DS receives its overall parcel mail volume by direct transportation from all input sorting centers. In this case, there is no assignment necessary to an OSC. Constraints (4) and (5) describe the flow conservation at sorting centers. In constraints (4), the incoming parcel mail volume at an ISC from the assigned areas and with unique DB/DS destination must equal the outgoing volume. The outgoing volume is described by the right-hand side of the equation and can be transported through long-haul or direct mode. In constraints (5), the incoming parcel mail volume at an OSC from all input sorting centers with unique DB/DS destination must equal the outgoing volume to the same DB/DS. Moreover, constraints (6) ensure that the overall parcel mail volume of a DB/DS is satisfied by transport flows from the input and output sorting centers. While constraints (7) and (9) imply that a transport flow is only possible if the corresponding assignment exists ($M = \infty$), constraints (8) allow a direct transport flow from an ISC to a DB/DS if it is bigger or equal to a minimum flow $N > 0$. Capacity constraints of the open input/output sorting centers are satisfied through inequalities (10) and (11). The number of open input/output sorting centers is restricted between a given minimum and maximum number through inequalities (11) and (12). Finally, constraints (13) and (14) state the binary or positive continuous nature of the decision variables.

Table 1 Number of variables of the SC model compared to the number of variables based on the aggregation of areas, the focus on delivery stations with guideline roles, and the restriction of the area assignment to the 7 nearest input sorting centers (Preprocessing)

Variable	Model	Preprocessing
x_{si}	271,722	4,676
y_{jt}	108,075	3,283
z_{it}	108,075	15,477
u_i	33	33
v_j	33	33
f_{ij}^t	3,566,475	108,339
f_{it}	108,075	15,472
f_{jt}	108,075	3,283
Total	4,270,563	150,596

2.2 Scenarios and Results

In the parcel distribution network being considered, 8,234 5-digit zip-code areas, 33 input/output sorting centers, 230 delivery bases (including 27 international European bases), and 3,075 delivery stations (for combined delivery of parcels and letters) exist. This numbers lead to more than 4 million decision variables and 500,000 constraints in the model of Sect. 2.1. For reasons of simplification and proportionality, we aggregated the areas by the first three digits of the 5-digit zip-code. Therefore, only 668 areas must be considered. Instead of 3,075 delivery stations, we focused on those with guideline roles (leading function for up to 15 delivery stations) as representatives, and reduced the number to 246. Further, we restrict the assignment of the areas to 7 nearest input sorting centers. In the same way, we restrict the assignment of the delivery bases/stations to the 7 nearest output sorting centers. We observed that more than 7 nearest sorting centers did not improve the solution. Under these assumptions and restrictions, defined together with Deutsche Post DHL, the number of decision variables and constraints of the model could be reduced drastically (see Tables 1 and 2). The model finally was solved to optimality with CPLEX 12.4 in acceptable time. Therefore, it was possible to analyze several different scenarios. Some of these scenarios will now be introduced and the according results will be presented.

In our scenarios we do not distinguish between input and output sorting centers ($I = J$). Therefore, we can omit one of the binary variables u_i or v_j and constraints (12) or (13). The following basis scenarios have been investigated among many others with different sets of costs and capacity parameters:

Scenario 1 Our first scenario (called "baseline scenario") represents the current parcel distribution network. In this scenario, all 33 location variables are fixed to the value '1' and the assignment variables are fixed according to the assignment of the current network. Hence, we only decide about the transportation flows of 2.8 million parcels of a representative day in 2012. All potential locations have the same fixed costs and the

Table 2 Number of constraints of the model compared to the number of constraints based on the aggregation of areas, the focus on delivery station with guideline roles, and the restriction of the area assignment to the 7 nearest input sorting centers (Preprocessing)

Constraints	Model	Preprocessing
(1)	8,234	668
(2)	3,275	469
(3)	108,075	15,477
(4)	108,075	3,283
(5)	3,275	469
(6)	108,075	15,477
(7)	108,075	15,477
(8)	108,075	15,477
(9)	33	33
(10)	33	33
(11)	2	2
(12)	2	2
Total	555,229	66,867

current sorting capacity. Direct transportation flows are omitted. The resulting objective value is used for comparisons.

Scenario 2 In addition to scenario 1, a new different assignment is allowed.

Scenario 3 In addition to scenario 2, the location and the number of sorting centers are not fixed.

Scenario 4 In addition to scenario 3, further potential sorting center locations are available (a total of 80 potential locations).

Scenario 5 Same as scenario 2, but direct transports to delivery bases/stations are allowed if the minimum flow exceeds N.

Scenario 6 Same as scenario 3, but direct transports to delivery bases/stations are allowed if the minimum flow exceeds N.

We now compare the model costs of the baseline scenario with the other described scenarios. First of all, all scenarios reduce the model costs (see Table 3). Even a new assignment of areas in the present situation leads to a reduction of approximately 2 % (scenario 2). The results of scenario 3 shows that a significant model costs reduction close to 5 % is possible. This is based on the fact that some of the sorting centers are not necessary in the model to handle the overall parcel mail volume from the areas to the delivery bases/stations. Considerations of including direct transports also slightly improve the model costs reduction (see Table 3, scenarios 5 and 6).

In another investigation, we fixed the number of open sorting centers of the baseline scenario to different values using constraints (12) with $Min_{\mathrm{ISC}} = Max_{\mathrm{ISC}} = Min_{\mathrm{OSC}} = Max_{\mathrm{OSC}}$ and assumed an uncapacitated case. Additionally, we repeated the investigation of the costs for the increasing parcel mail quantities of 25, 50, and 100 %, respectively, to the baseline scenario. All model costs curves for the different assumed parcel mail volume look similar (see Fig. 4). Obviously, model costs do not have a one-to-one relationship to the parcel mail volume. Moreover, the model costs

Table 3 Scenario 1 is our baseline scenario. Model costs of scenarios 2–5 are compared with scenario 1, and the reduction is given in percent

Scenario	Model costs reduction in (%)	Direct transport
1	0.00	no
2	1.98	no
3	5.08	no
4	5.61	no
5	2.75	yes
6	6.12	yes

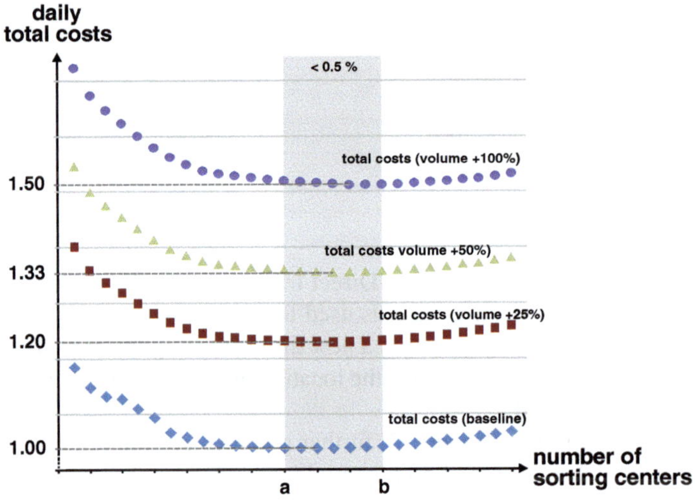

Fig. 4 Model costs curves for different number of sorting centers and scenarios considering a future increase in parcel mail volume. The best solution of the baseline scenario is set to the reference value 1.0

only differ less than 0.5 % within a range [a, b] number of sorting centers for all four volume scenarios (see Fig. 4, depicted gray area).

Altogether, the results show that scenarios with direct transportation mode, modified geographical assignment and fewer sorting centers, respectively, are more cost efficient than the current parcel distribution network.

3 Long-haul Transportation in the Letter Mail Network

As stated earlier, one major difference between the long-haul transports in the parcel and letter mail transportation networks is the use of hubs. While no hubs are used in the German parcel mail network because of full truckloads between the sorting centers, hubs are used in the German letter mail network, since letters are much smaller than parcels, resulting in less-than truckload shipments between sorting centers despite the daily letter mail volume being far greater than the parcel one. Therefore, in order to reduce transportation costs, carriers have to exploit economies of scale through the consolidation of flows between the sorting centers by using hubs.

In addition to solving a location/allocation model like the one described above, it is also necessary to decide upon the location of the hubs used for the long-haul transports and the allocation of the sorting centers to the hubs. That kind of decision, i.e. the design of a hub-and-spoke network, is usually made by solving so-called hub location problems, which have been the subject of numerous works. We refer to [7]; and [2] for a comprehensive review and to [10] for an insight to the current status of this research field.

Economies of scale achieved through consolidation of flows is the *raison d'être* for hub-and-spoke networks and is usually modeled by discounting the unit costs of transportation for inter-hub flows with a discount factor $0 < \alpha < 1$ to reflect the consolidation of flows between hub nodes. Let $G = (N, A)$ be a complete graph, where $N = \{1, \ldots, n\}$ denotes the set of sorting centers and potential hub location respectively, then c_{ij} indicates the unit costs of transportation between the nodes i and j for every ordered pair $(i, j) \in N \times N$. Assuming that flows between sorting centers i and j go through paths $i \to k \to l \to j$, where k and l are hub nodes, the total unit costs of transportation can be expressed by $c_{ijkl} = c_{ik} + \alpha c_{kl} + c_{lj}$ with $0 < \alpha < 1$. While this approach is widely spread in the literature because of its simplicity, it has been facing substantial criticism e.g. [9, 25].

This criticism mainly addresses the fact that the discount factor α is flow-independent, meaning that it does not depend on the actual flow on the hub arcs. This is a mismatch between the model and the underlying idea that economies of scale are achieved through consolidation of flows on arcs with high flows, since it often leads to optimal solutions in which some hub arcs carry considerably less flow than most of the non-hub arcs [9]. As a consequence, there is no reason why economies of scale should apply to hub arcs only. Furthermore, it is not clear what value should be used for α. Values used in the literature range from 0.25 [16] up to 0.7–1.0 [13]. Another issue with this approach to modelling economies of scale is the fact that the number of hubs needs to be determined in some way in the corresponding models. This is either done by predefining the number p of hubs to be installed in so-called p-hub Median problems [8] or by assuming that installing a hub in a node k of the network incurs fixed costs F_k in hub location problems [30]. Some authors have tried to avoid the above mentioned criticisms and introduced flow-dependent discount factors, resulting in non-linear objective functions which can be approximated by piecewise linear functions e.g. [5, 27, 29]. In a simpler approach, [32] consider flow thresholds for every hub-arc. These thresholds must be reached in order for the unit costs of transportation on a particular arc to be discounted [9] introduced the so-called hub arc location problem, where hub arcs are explicitly selected instead of hub nodes. This usually leads to solutions with hub arcs carrying more flow than in the corresponding hub location problems.

Even if the approaches mentioned above do provide considerable improvements to the way economies of scale are modeled in hub location models, some of the major criticisms mentioned earlier still apply. In all of the works mentioned above,

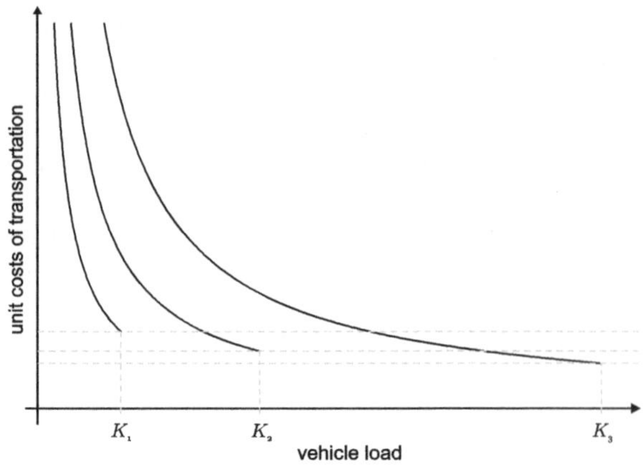

Fig. 5 Unit costs of transportation for different vehicle capacities as a function of the vehicle load

economies of scale still apply to hub arcs only, and the number of hubs still needs to be restricted in some way. To resolve these problems, we have developed a new approach to economies of scale in hub-and-spoke networks, which is trip-based. While in all of the works mentioned above, transportation costs are always incurred by flows between nodes, we chose an approach in which transportation costs are incurred by vehicle trips. This approach has two advantages. If we assume fixed costs $c_f^{fix} = c_f$ for a specific vehicle trip f, then the unit costs of transportation as a function of the vehicles load l_f is given by

$$\frac{c_f^{fix}}{l_f}, \text{ if } l_f > 0$$

$$0, \text{ otherwise,}$$

which is a very simple but adequate description of economies of scale illustrated in Fig. 5 for different vehicle capacities K_1, K_2, and K_3.

The second advantage of this approach is the fact that it allows easy consideration of different kinds of consolidation. Traditional hub location models only consider the consolidation of items with different origins and destinations on hub arcs, even though, according to [12], there are several ways in which items can be consolidated. The consolidation of items with the same origin and different destinations or with different origins and the same destination, as illustrated in Fig. 6, is of great interest and relevance when optimizing long-haul transports in letter mail networks, since it is a very effective way to reduce transportation costs.

3.1 Problem and Formulation

The main focus of this section is to find the optimal number and locations of sorting centers to be used as hubs in the long-haul transports in Deutsche Post DHL's letter

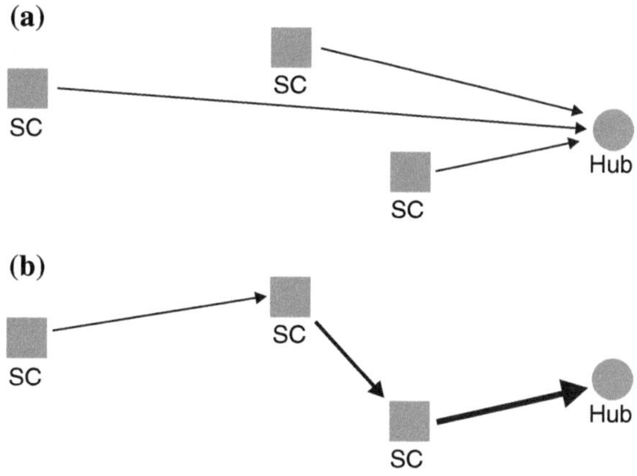

Fig. 6 Consolidation of items with different origins and the same destination according to [12], **a** without consolidation. **b** with consolidation

mail transportation network. This problem in fact is, if we look at the long-haul subnetwork only, a hub location problem. However, one distinctive feature of this problem is the absence of fixed costs for using a sorting center as a hub and the desire not to restrict the number of hubs in any way. Because of these characteristics, in order to optimize, we used the above mentioned idea to consider costs for vehicle trips instead of costs for flows, since it does not require the number of hubs to be restricted in any way in order to provide meaningful solutions. Therefore, as a part of the above-mentioned hub location problem, we also need to generate a set of cost-minimal vehicle trips and to allocate flows to these vehicle trips, ensuring that all items are shipped from their origin to their destination.

Another peculiarity of the problem studied is the fact that the hub nodes are not capacitated. This assumption is realistic, since all the sorting takes place in the inbound sorting centers (see Fig. 1) and the items are only transshipped in the hubs. However, the sorting capacities in the outbound sorting centers are restricted and decrease with time. For the sake of simplicity, the sorting capacities were discretized and are checked at regular intervals (see Fig. 7).

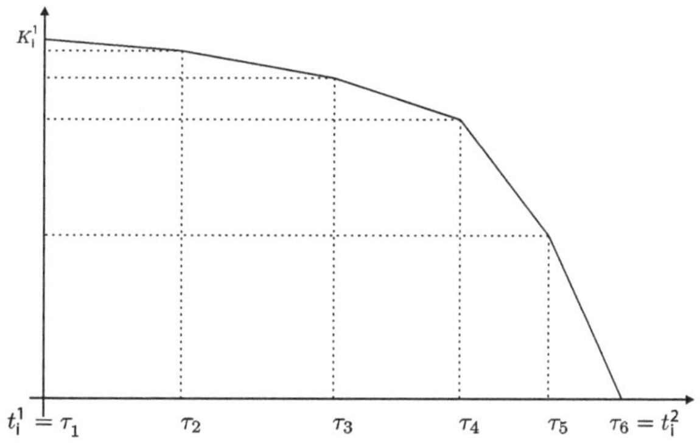

Fig. 7 Discretized time-varying sorting capacity function according to [22]

Consider the complete graph $G = (N, E)$, where $N = \{1, \ldots, n\}$ represents the set of nodes corresponding to origins/destinations as well as potential hub locations and E is the set of edges. Let a_{ij} denote a transportation request from $i \in N$ to node $j \in N$, and m_{ij} denote the corresponding letter mail volume from node $i \in N$ to node $j \in N$, t^a_{ij} the time when the letter mail volume becomes available at node i, and l_{ijkl} a path from node $i \in N$ to node $j \in N$ via hubs k and $l \in N$. For every node i, let t^1_i denote the opening time and t^2_i the closing time of node $i \in N$. Because of the time window $\left[t^1_i, t^2_i\right]$ in every node of the network, not every path l_{ijkl} is feasible. A path is only feasible if the time windows defined by the earliest arrival time and latest departure time and the time window of the respective nodes overlap for every node on the path. We define the set H_{ij} to contain all combinations (k, l) of hubs, such that the resulting path l_{ijkl} is feasible with regard to time.

The above-mentioned path l_{ijkl} only gives a rough idea of how the flow is actually routed through the network. This depends on how the flows on a specific path l_{ijkl} are allocated to vehicle trips. Formally, a vehicle trip $f \in F$ corresponds to an alternating sequence of nodes and arcs on the graph G, where F denotes the set of all vehicle trips. Let $\mathscr{S}_f$ denote the set of all nodes and $\mathscr{P}_f$ the set of all arcs for a specific vehicle trip $f \in F$, $t^1_{f,s}$ the arrival time of trip f in node s, and $t^2_{f,s}$ the corresponding departure time. The path l_{ijkl} for transportation request a_{ij} can now be split into three distinct transportation requests $\bar{a}^1_{ik}$, $\bar{a}^2_{kl}$, and $\bar{a}^3_{lj}$, which need to be fully allocated to some vehicle trips $f \in F$. For this we need to make sure that the requests $\bar{a}^1_{ik}$, $\bar{a}^2_{kl}$, and $\bar{a}^3_{lj}$ are compatible, with regard to time, with the vehicle trips they are allocated to. By $\mathscr{F}_{\bar{a}^1_{ik}}$, $\mathscr{F}_{\bar{a}^2_{kl}}$, and $\mathscr{F}_{\bar{a}^3_{lj}}$ we denote the sets of vehicle trips $f \in F$ compatible with $\bar{a}^1_{ik}$, $\bar{a}^2_{kl}$, and $\bar{a}^3_{lj}$, meaning that the corresponding arcs (i, k), (k, l), or (l, j) need to be elements of $\mathscr{P}_f$ and the time windows $[t^1_{f,i}, t^2_{f,i}]$, $[t^1_{f,k}, t^2_{f,k}]$ etc. need to overlap with the earliest arrival time and latest departure time for the transportation requests in the nodes i, k, l, and j. The set $\mathscr{A}$ contains all transportation requests $\bar{a}$ where $\bar{m}$ represents the corresponding volume. Finally $\mathscr{E}_j$ denotes the set of all transportation requests ending in node $j \in N$, $\mathscr{E}^{\mathscr{F}}_j$ the set of vehicle trips compatible with the aforementioned transportation requests, and $\mathscr{E}^{\mathscr{F},\tau}_j$ the subset of $\mathscr{E}^{\mathscr{F}}_j$ containing all vehicle trips arriving in node j after time τ.

$$\min z = \sum_{f \in F} c_f \cdot z_f + \sum_{k \in B} c_{Hub} \cdot y_k \tag{16}$$

subject to

$$\sum_{(k,l) \in H_{ij}} x_{ijkl} = 1, \quad \forall i, j \in N \tag{17}$$

$$\sum_{l \in N} x_{ijkl} \leq y_k, \quad \forall i, j, k \in N \tag{18}$$

$$\sum_{l \in N} x_{ijlk} \leq y_k, \quad \forall i, j, k \in N \tag{19}$$

$$\sum_{f \in \mathscr{F}_{\bar{a}}} z_{\bar{a}}^{f} = 1, \quad \forall \bar{a} \in \mathscr{A} \tag{20}$$

$$z_{\bar{a}}^{f} \leq z_f, \forall f \in \mathscr{F}, \forall \bar{a} \in \mathscr{A} \tag{21}$$

$$\sum_{\bar{a} \in \mathscr{A}} q_{f,p}^{\bar{a}} \cdot z_{\bar{a}}^{f} \cdot \bar{m} \leq Q_f, \forall f \in \mathscr{F}, \forall p \in \mathscr{P}(f) \tag{22}$$

$$\sum_{\bar{a} \in \mathscr{E}_j} \sum_{f \in \mathscr{E}_j^{\mathscr{F},\tau}} z_{\bar{a}}^{f} \cdot \bar{m} \leq K_j^{\tau}, \forall j \in N, \forall \tau \in [t_j^1; t_j^2] \tag{23}$$

$$y_k \in \{0, 1\} \, \forall k \in N \tag{24}$$

$$z_f \in \{0, 1\} \, \forall f \in \mathscr{F} \tag{25}$$

$$x_{ijkl} \in [0, 1] \, \forall i, j, k, l \in N \tag{26}$$

$$z_{\bar{a}}^{f} \in [0, 1] \, \forall \bar{a} \in \mathscr{A}, \forall f \in \mathscr{F} \tag{27}$$

The routing variables $x_{ijkl} \in [0, 1]$ represent the fraction of volume m_{ij} on the corresponding path l_{ijkl}. The binary variables $z_f \in \{0, 1\}$ are equal to 1 if vehicle trip $f \in F$ is realized, and 0 otherwise, while the continuous variables $z_{\bar{a}}^{f} \in [0, 1]$ represent the fraction of transportation request $\bar{a} \in \mathscr{A}$ allocated to vehicle trip $f \in F$. Finally, the binary variables $y_k \in \{0, 1\}$ are equal to 1 if node $k \in N$ may be used as a hub and equal to zero otherwise. The problem can then be formulated as follows:

The objective function (16) aims at minimizing the total costs consisting of fixed set-up costs for the located hubs and fixed transportation costs for the vehicle trips. Constraints (17) ensure that for every pair (i, j) the total volume m_{ij} is routed via some feasible hubs k and l. Constraints (18) and (19) state that nodes k and l may only be used as hubs if they are hub nodes. Constraints (20) guarantee that every transportation request is fully allocated to vehicle trips, while constraints (21) states that requests only can be allocated to vehicle trips that are realized. Constraints (22) impose capacity constraints on every vehicle trip. Constraints (23) ensure that capacity constraints hold for all destination nodes j at every (discrete) point in time. Finally, constraints (24)–(27) state the binary or continuous nature of the decision variables.

3.2 Solution Approach and Results

The model given in the previous section is a possible mathematical formulation for the problem described above. When trying to optimize the long-haul transports between the 82 sorting centers in Deutsche Post DHL's letter mail transportation network, the problem very quickly becomes computationally intractable for commercial MIP solvers. Hence, the problem was solved by means of a two-stage heuristic approach. We start by constructing a feasible solution which is then improved by a tabu search procedure. Tabu search is a metaheuristic solution technique which has proven to be very successful in solving optimizations problems with a complex solution space. For details of tabu search, we refer to [17–19].

3.2.1 Construction Heuristic

In order to construct a first feasible solution, we identify arcs of the network which offer a high potential for consolidation. This is done by computing all paths feasible with regard to time for every origin-destination relation in the long-haul transportation network and then calculating the maximum flow compatible with regard to time on every arc of the network. The arc with the maximal compatible flow is then chosen and used to generate a new vehicle trip between the corresponding nodes.

During this procedure we allocate the maximum flow possible to the new trip, where one can chose between vehicles with different capacities as part of the procedure. Alternatively, the maximum flow also can be allocated to an existing tour which is extended and/or altered during this process if necessary.

Eventually, the flows need to be adjusted for all of the origin-destination relations allocated in the current iteration. This adjustment not only affects the flow on the chosen arc but the flows on the arcs of all alternative feasible paths as well. This procedure is repeated until all flows have been allocated to vehicle trips and feasibility for the above mentioned problem is hereby guaranteed. The construction heuristic can be summarized as follows:

Step 1 For each origin-destination relation (i, j), compute all feasible path from i to j via a maximum of two hubs which are feasible with regard to time

Step 2 For every arc of the network, calculate the maximum flow compatible with regard to time windows

Step 3 Find the arc with the highest maximum compatible flow

Step 4 Create a new or extend an existing vehicle trip and allocate as much flow as possible

Step 5 Adjust the flows for all origin-destination relations allocated in step 4 on every feasible path computed in step 1

Step 6 Repeat Steps 1 to 5 until all flows are allocated

3.2.2 Tabu Search

In the second stage, this first feasible solution is then improved by means of a tabu search procedure. For this we defined several neighborhood move procedures which can be categorized into two groups. Procedures from the first group deal with location decisions concerning the hubs while those from the second group address the vehicle trips.

Add/Drop	Starting with a set of nodes containing the current potential hub nodes, this procedure adds a new or removes an existing potential hub node to/from the aforementioned set.
Swap	This procedure adds a node to the set of potential hub nodes if it is currently not contained in the set and removes it from the set if it currently is among the set's elements.
Merge	Merges two existing vehicle trips if the resulting trip remains feasible with regard to capacity and time. The vehicle's capacity can be increased as a part of the process.
Increase/Decrease capacity	Increases or decreases the capacity of the vehicle for a specific trip if the trip's feasibility is not affected. Feasibility can be affected by an insufficient capacity of the corresponding vehicle or delays in the vehicle's trip resulting from vehicle with a higher capacity being slower.
Remove	Removes an existing vehicle trip if it does not currently have any flows allocated to it or if these can be allocated to existing vehicle trips.
Swap request	This procedure de-allocates a transportation request from a vehicle trip and allocates it to an existing or a new vehicle trip. This can increase the original trip's flexibility with regard to time and therefore allow other moves which were previously infeasible.

These local search procedures are embedded in and guided by a tabu search procedure. Our observation was that the neighborhood is too large for all its improving solutions to be generated and identified. We therefore opted for a k best search probabilistic tabu search [33], which randomly selects one of the above mentioned neighborhood move procedures, where all procedures share the same probability of being selected. In every iteration of the tabu search procedure, the best improving move among the k best is selected and the procedure is repeated until a maximum number of iterations or a maximum number of iterations with no improving neighborhood move is reached.

3.2.3 Results

The proposed two-stage heuristic solution procedure described above was coded in C++ and executed on an Intel® Xeon® X5680 3.3GHz CPU with 48GB of RAM. It was used to optimize the long-haul transports in Deutsche Post DHL's letter mail transportation network. This network features a total of 82 sorting centers resulting in a total of 6,642 relations to be considered.

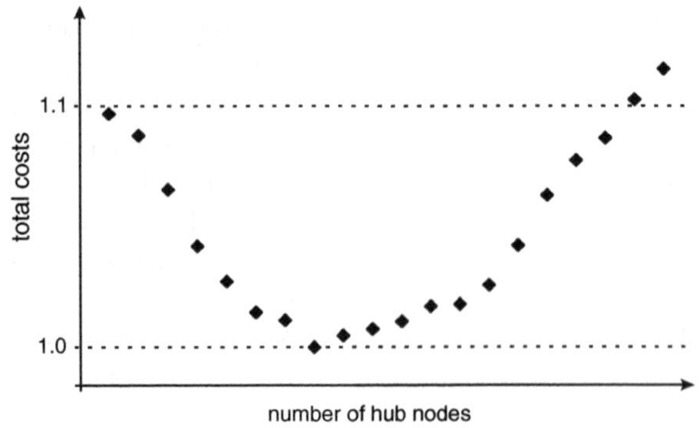

Fig. 8 Total transportation costs for the best solutions obtained for different numbers of hub nodes in the long-haul network of Deutsche Post DHL's letter mail network

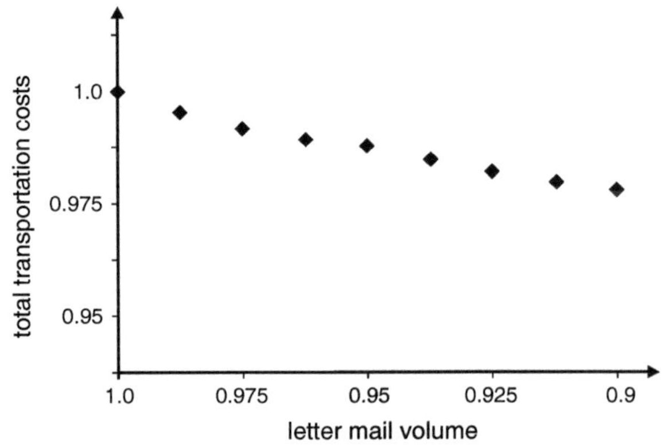

Fig. 9 Total transportation costs for the best solutions obtained for decreasing letter mail volume

In the so-called baseline scenario we considered present-day volumes and the corresponding optimization led to a substantial cost reduction with only minor changes to the long-haul network's hub configuration. Most notably though, despite only considering transportation costs and not assuming any fixed costs for hub location, the optimization showed that it is not optimal to use all nodes of the network, as hub nodes as the classical hub location models mentioned above would suggest (c.f. Fig. 8).

We considered other different scenarios taking into account a decrease in letter mail volume of up to 10 %. As can be seen in Fig. 9, total transportation costs decrease with decreasing letter mail volume, but to a lesser extent. That is, a decrease in letter mail volume of 10 % leads to a decrease in total transportation costs of less than 2.5 %. This can be explained by the assumption that costs for vehicle trips are fixed and do not depend on the vehicle load. Therefore, total transportation costs can only be reduced by reducing the number of vehicle trips required to transport all of the letter mail. This is in sharp contrast to the classical models, where transportation costs are a linear function of the volume, and a decrease in letter mail volume leads to an equivalent drop in transportation costs.

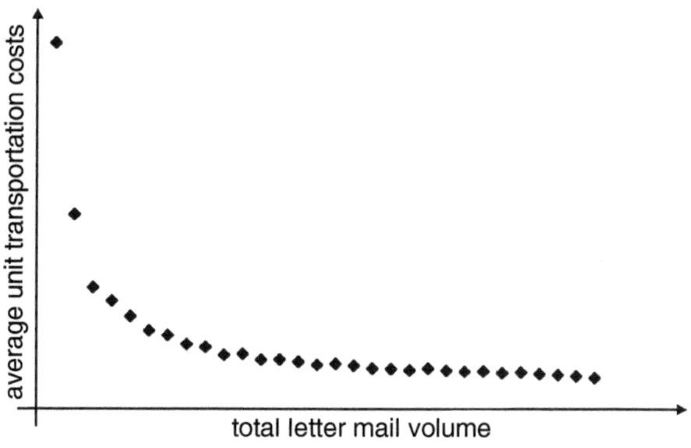

Fig. 10 Average unit costs of transportation for increasing total letter mail volume

This issue is also reflected in Fig. 10, which depicts the impact of increasing total letter mail volume on average unit costs of transportation in the long-haul transportation network. The relationship is clearly not linear and unit costs of transportation decrease asymptotically for increasing letter mail volume, as suggested in Fig. 5.

Altogether, the above results show that the chosen approach for modelling economies of scale is adequate and provides meaningful results. Furthermore, our approach also revealed potential to substantially reduce the total transportation costs in Deutsche Post DHL's long-haul transportation network.

4 Conclusion

The analysis of the results for the parcel mail network shows that solutions with direct transportation mode, modified geographical assignment and less sorting centers, respectively, are cost efficient. However, we do not consider any service-level aspects in our strategic model. In this context, the so called E + 1 ratio (percentage of parcels, which reach their destination within the next day) is a key performance indicator for quality. Currently, on average 85 % of all parcels reach their recipients the next day. A post analysis revealed that all scenarios could fulfill the current E+1 ratio.

In future, it is proposed that the E+1 ratio will be increased to 95 % by 2022. Therefore, we formulated an extended strategic MILP model with transportation times to perform analyses on the E+1 ratio. However, only small instances can currently be solved to optimality with CPLEX because of the high degree of complexity. To cope with real-world instances, a heuristic approach based on tabu search has been developed and is currently under investigation. First investigations show very promising results. Hence, quality aspects also should be included in the location-allocation process. The results of these investigations will be presented in follow-up scientific publications.

As far as the letter mail network is concerned, our results reveal substantial optimization potential in the long-haul transportation network and also support the soundness of the newly developed approach to model economies of scale in hub location models in an adequate way. In future, we plan to integrate service quality constraints into the model as well in order to be able to analyze the impact of quality requirements on transportation costs and hub location decisions.

Currently, we assume that all the letter mail volume has to reach destination the next day (i.e. a E+1 ratio of 100 %), which is very expensive in terms of transportation costs because of the time window given for the long-haul transportation. Hence, transportation costs could be significantly reduced by hypothetically allowing E+1 ratios of under 100 %. These considerations will be subject of future work.

References

1. AIMMS: Paragon decision technology (2012). http://www.aimms.com
2. Alumur S, Kara B (2008) Network hub location problems: the state of the art. Eur J Oper Res 190(1):1–21
3. Armacost AP, Barnhart C, Ware K, Wilson A (2004) Ups optimizes air its network interfaces 34(1):15–25
4. Bruns A, Klose A, Stähly P (2000) Restructuring of swiss parcel delivery services. OR Spectr 22:285–302
5. Bryan D (1998) Extensions to the hub location problem: formulations and numerical examples. Geogr Anal 30(4):315–330
6. Büdenbender K, Grünert T, Sebastian HJ (2000) A hybrid tabu search/branch-and-bound algorithm for the direct flight network design problem. Transp Sci 34(4):364–380
7. Campbell AM, Ernst AT, Krishnamoorthy M (2002) Hub location problems, chapter hub location problems. Springer, Berlin, pp 373–407
8. Campbell JF (1996) Hub location and the p-hub median problem. Oper Res 44(6):923–935
9. Campbell JF, Ernst AT, Krishnamoorthy M (2005) Hub arc location problems: part i-introduction and results. Manag Sci 51(10):1540–1555
10. Campbell JF, O'Kelly ME (2012) Twenty-five years of hub location research. Transp Sci 46(2):153–169
11. CPLEX: Ibm ilog (2012). http://www-01.ibm.com/software/integration/optimization/cplex-optimizer/
12. Crainic TG (2000) Service network design in freight transportation. Eur J Oper Res 122(2):272–288
13. Cunha CB, Silva MR (2007) A genetic algorithm for the problem of configuring a hub-and-spoke network for a LTL trucking company in Brazil. Eur J Oper Res 179:747–758
14. DHL: Press release (2011). http://www.dp-dhl.com/en/media_relations/press_releases/2011/deutsche_postdhl_expands_parcel_network.html
15. Ernst AT, Hamacher H, Jiang H, Krishnamoorthy M, Woeginger GJ (2009) Uncapacitated single and multiple allocation p-hub center problems. Comput Oper Res 36(7):2230–2241
16. Ernst AT, Krishnamoorthy M (1996) Efficient algorithms for the uncapacitated single allocation p-hub median problem. Loc Sci 4:139–154
17. Glover F (1989) Tabu search - part i. ORSA J Comput 1(3):190–206
18. Glover F (1990) Tabu search - part ii. ORSA J Comput 2(1):4–32
19. Glover F, Laguna M (1997) Tabu search. Kluwer Academic Publishers, Norwell
20. Grünert T, Sebastian HJ (2000) Planning models for long-haul operations of postal and express shipment companies. Eur J Oper Res 122(2):289–309
21. Gündüz, HI (2011) The single-stage location-routing problem with time windows. In: Böse J, Hu H, Jahn C, Shi X, Stahlbock R, Voß S (eds) ICCL. Lecture Notes in Computer Science, vol 6971. Springer pp 44–58

22. Hempsch C, Irnich S (2008) Vehicle routing problems with inter-tour resource constraints. Operations Research/Computer Science Interfaces Series, vol 43. Springer, Berlin, pp 421–444
23. Irnich S (2002) Netzwerk-design für zweistufige transportsysteme und ein branch-and-price-verfahren für das gemischte direkt- und hubflugproblem. Ph.D. thesis, Lehr- und Forschungsgebiet Operations Research und Logistik Management, RWTH Aachen University, Aachen, Germany
24. Kara BY, Tansel BC (2000) On the single-assignment p-hub center problem. Eur J Oper Res 125(3):648–655
25. Kimms A (2006) Economies of scale in hub and spoke network design models: we have it all wrong. Gabler Edition Wissenschaft. Deutscher Universitäts-Verlag, Wiesbaden, pp 293–317
26. Klincewicz JG (1998) Hub location in backbone/tributary network design: a review. Loc Sci 6(1–4):307–335
27. Klincewicz JG (2002) Enumeration and search procedures for a hub location problem with economies of scale. Ann Oper Res 110:107–122
28. Lischak C (2001) Standortplanung für einen privaten paketdienstleister. Ph.D. thesis, Fakultär füt Mathematik, Informatik und Naturwissenschaften, RWTH Aachen University, Aachen, Germany
29. O'Kelly M, Bryan D (1998) Hub location with flow economies of scale. Transp Res B-Methodol 32(8):605–616
30. O'Kelly ME (1992) Hub facility location with fixed costs. Pap Reg Sci 71:293–306
31. Pajunas A, Matto EJ, Trick M, Zuluaga LF (2007) Optimizing highway transportation at the United States postal service. Interfaces 37(6):515–525
32. Podnar H, Skorin-Kapov J, Skorin-Kapov D (2002) Network cost minimization using threshold-based discounting. Eur J Oper Res 137:371–386
33. Rochat Y, Taillard ED (1995) Probabilistic diversification and intensification in local search for vehicle routing. J Heuristics 1:147–167
34. Sebastian HJ (2012) Optimization approaches in the strategic and tactical planning of networks for letter, parcel and freight mail. Lecture Notes in Business Information Processing, vol 42. Springer, Berlin, pp 36–61
35. Sebastian HJ (2012) Recent advances in strategic and tactical planning of optimal postal networks. Presented at the INFORMS conference on business analytics and operations research. Huntington Beach
36. Wasner M, Zäpfel G (2004) An integrated multi-depot hub-location vehicle routing model for network planning of parcel service. Int J Prod Econ 90(3):403–419
37. Winkelkotte TJ (2010) Strategische optimierung von distributionsnetzwerken - ein optimierungsmodell und heuristische lösungsverfahren zur planung von standorten und absatzgebieten mit approximativer berücksichtigung der taktischen und operativen logistikprozesse. Ph.D. thesis, Deutsche Post Lehrstuhl für Optimierung von Distributionsnetzwerken, RWTH Aachen University, Aachen, Germany
38. Wollenweber JG (2007) Mehrstufige facility-location-probleme mit stückweise linearen kosten: Modellierung und flexible heuristische lösungsverfahren. Ph.D. thesis, Deutsche Post Lehrstuhl für Optimierung von Distributionsnetzwerken, RWTH Aachen University, Aachen, Germany

Recent Advances in Strategic and Tactical Planning of Distribution Subnetworks for Letter Mail

Halil Ibrahim Gündüz, Christoph Klemens Hempsch and Hans-Jürgen Sebastian

Abstract This paper considers the postal logistics area, more precisely, the distribution networks for letter mail. A main service provided by postal companies is letter mail transportation and delivery. In this market segment there have been two key efforts during the last few years: reduction in transportation and delivery time (service quality) and minimization of costs under service quality constraints. Both efforts—reduction of service time and minimization of costs for providing the promised services—have a strong impact on the quality of the strategic and the tactical planning phases of the respective distribution networks. The Operations Research type of analytical models used in the strategic and tactical planning phases of distribution networks in postal organizations are: facility location, location routing, service networks design, and vehicle routing and scheduling models. In this article we introduce the structure of a typical distribution network for letter mail and for parcel mail, and we describe the main subnetworks. This paper is also concerned with projects on optimization of such subnetworks. Therefore, we have selected three projects dealing with different subsystems and covering the strategic and the tactical planning phases as well. The projects are in the areas of collecting mail from mailboxes (vehicle routing), replanning of delivery station locations (facility location combined with vehicle routing), and reducing deadheading in the last mile (facility location combined with vehicle routing). Each of the projects covers system analysis, modeling, development of optimization algorithms, and a software prototype.

H.I. Gündüz (✉) · H. Sebastian
Deutsche Post Chair of Optimization of Distribution Networks,
RWTH Aachen University, Kackertstr. 7, 52072 Aachen, Germany
e-mail: guenduez@dpor.rwth-aachen.de

H. Sebastian
e-mail: sebastian@dpor.rwth-aachen.de

C.K. Hempsch
Deutsche Post AG, Charles-de-Gaulle-Str. 20, 53113 Bonn, Germany
e-mail: hempsch@deutschepost.de

H.-J. Sebastian et al. (eds.), *Quantitative Approaches in Logistics and Supply Chain Management*, Lecture Notes in Logistics,
DOI 10.1007/978-3-319-12856-6_5

1 Introduction

The increasing market competition and the service focus of customers force logistics service providers, such as postal organizations and express shipment companies, to evaluate and to continuously improve their networks for letter, parcel, and freight mail.

The key service provided by postal companies is letter mail and parcel mail transportation and delivery. In this market segment there have been two key efforts during the last few years:

- reduction in transportation and delivery time (service quality)
- minimization of costs under service quality constraints

In addition, the goal of 'green logistics' has become more and more important and is today firmly established as one of the core values of several leading postal service providers, such as Deutsche Post DHL.

Both efforts—the reduction of service time and the minimization of costs for providing the promised services—have a strong impact on the quality of the strategic and the tactical planning phases of the respective distribution networks. Instead of using simple techniques in order to get a *quick solution*, advanced model-based optimization and simulation are needed today.

Optimization of facility locations and the allocation of *customers* to the facilities, such as terminals, depots, sorting centers, and hubs, are the most important decisions within the strategic planning phase. If the facilities and their locations are selected and the allocation of *customers* is done simultaneously, the tactical planning phase includes the optimization of transportation and delivery (*the last mile*) in order to determine the routes and the schedules for the fleet of vehicles used in the distribution network. The Operations Research type of analytical models used in the strategic and tactical phases of planning of distribution networks in postal organizations are: facility location, location routing, service network design, and vehicle routing and scheduling models (problems).

1.1 The Distribution Networks for Letter Mail and Parcel Mail

In this section we introduce the structure of a typical distribution network for letter mail, parcel mail, or both types of mail together (see also [17]). In Fig. 1 such a network is shown, and numbers for sorting centers and delivery stations are mentioned that relate to the distribution network of Deutsche Post DHL for letter mail within Germany. This network can be considered to be composed of four main *stages* (subnetworks):

- Stage 0—mail collection: This subnetwork collects the mail from different mail sources, and uses consolidation points in order to transport the mail to the sorting centers.

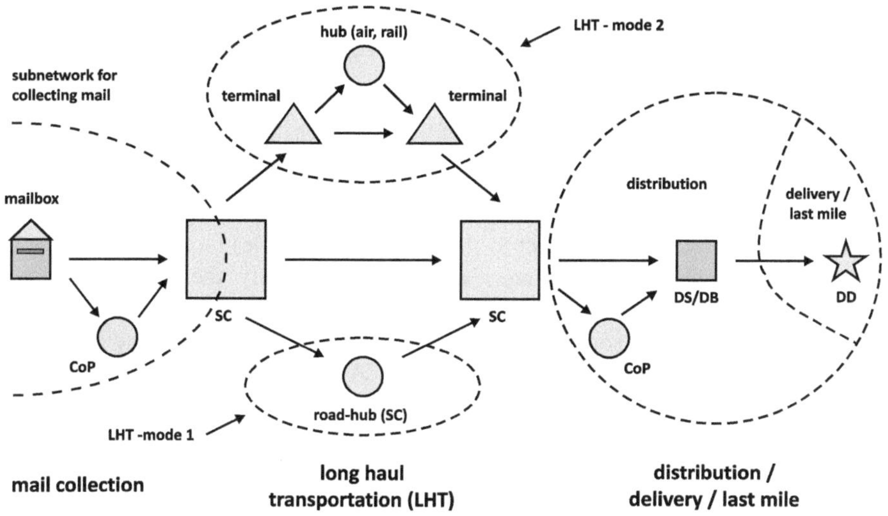

Fig. 1 Components/subnetworks of a distribution network for letter mail and parcel mail

- Stage 1—long haul transportation (LHT): This subnetwork realizes the exchange of mail between sorting centers. The idea is to use consolidation and bigger or faster vehicles for the long distances between sorting centers [5].
- Stage 2—distribution: These distribution networks (used in a more narrow sense) distribute the mail from sorting centers to mini-hubs (so-called delivery stations or delivery bases), where the final preparation for the postmen's tours takes place and where the postmen usually start their delivery routes (the last mile).
- Stage 3—delivery (last mile): Postmen visit their assigned delivery districts in order to deliver the mail to private households or to business customers.

If one were to compare such a distribution network in postal logistics with a classical distribution network for physical goods, several important differences would become evident (e.g. a small number of product types in the postal case, but a huge number of mail sources, mail final destinations, and commodity exchange processes within the same stage). We go slightly further into the details of the stages (subnetworks) described above in order to be able to characterize the special optimization projects that we will discuss in Sects. 2–4 in detail. First, we consider the subnetworks for collecting mail. Usually, each sorting center (SC) has its own network for mail collection. The main objects belonging to the mail collection networks are

- mail sources, e.g., mailboxes, business customers, retail stores (may also be used as consolidation terminals), and
- collection routes, which perform mail transportation from the mail sources to their allocated sorting centers.

Sorting centers for letter or parcel mail, respectively, are big automated facilities which implement the sorting part (the *production*) within these distribution networks.

Sorting centers for letter mail work in two different modes: SCA and SCE. The SCA mode performs the sorting of collected mail with respect to the destination SC (in Germany, the destination SC is characterized by the first two digits of the zip code). The SCE mode undertakes automated sorting of the incoming mail from the long-haul transportation network with respect to distribution (from the SC to the delivery stations) and with respect to the delivery (last mile) (sorting according to the sequence in which the postman visits his DD). Usually, the geographic areas allocated to the SCA mode and the SCE mode of an SC are identical. However, there is an option to use some of the sorting centers during several time periods of the year as SCE centers only. Physically, the SCA and SCE sorting centers are the same. An SC can operate either in SCA or in SCE mode, depending on time of day. This is possible because of the high degree of flexibility of the automatized sorting machines.

Long haul transportation means the global area transportation network which connects the sorting centers with each other. Consolidation is used in order to transport big quantities of mail over longer distances using larger vehicles or using multi-modal transportation (e.g. road-air, road-rail).

Finally, there are subnetworks for distributing the mail from an SC (working in SCE mode) to the *delivery stations* and the so-called *last mile*. A delivery station (DS) is a mini-hub, where the final preparation for delivery takes place as performed by the postmen. Then, the postmen pick up their sorted mail and start the delivery process, each of them visiting their assigned delivery district (DD) by car, by bicycle, or on foot in a predefined (ideally optimized) sequence.

Figure 1 shows a more schematic picture of a distribution network for letter mail and parcel mail. The subnetworks and their components are denoted by the abbreviations introduced before. In order to illustrate the dimension of an instance of such a distribution network, we give some characteristic approximate numbers of Deutsche Post DHL distribution network for letter mail within Germany:

- 40 million final destinations, including 3 million business customers
- approximately 68 million letters (of different types) every working day
- 1,08,000 mailboxes
- 82 sorting centers plus the international postal center in Frankfurt
- 3,100 delivery stations and 14,000 offices (retail)
- 53,000 delivery districts (3,500 visited on foot, 18,500 by bicycle, and 31,000 by car)

There are approximately 80,000 postmen employed by Deutsche Post DHL within the last mile in Germany.

1.2 Strategic and Tactical Planning of Subnetworks

The huge size of the networks considered above does not allow the development of an overall optimization model which has a chance of being solved either exactly or approximately. Also, the acquisition of all data needed as an input for such a model seems to be either impossible or much too demanding in terms of time and costs. Therefore, in order to reduce complexity, the well known planning phases are introduced. In addition, the overall network is heuristically decomposed into subnetworks by introducing an overall service-quality level for the whole network (e.g., next-day delivery for all postal products) and by assigning time windows and cut-off-times to the predefined subnetworks (see Fig. 1), such that the overall service quality can be fulfilled. This approach is described in more detail in [16]. For example, the long-haul transportation network for letter mail in Germany has a time window on weekdays from 9 p.m. to 4 a.m. The cut-off time for the mail collection network is 9 p.m. Comparable standards are found in [14] for transportation analysis and cost-savings opportunities in the surface-transportation network at the United States Postal Service. In the past, the Deutsche Post Chair of Optimization of Distribution Networks at RWTH Aachen University and its predecessors have successfully executed a number of projects dealing with the optimization of subnetworks of the distribution network in postal logistics, e.g.,

- the optimization of the Deutsche Post Night-Airmail network for letter mail (LHT subnetwork for letter mail) [3, 4, 8]
- the delivery station location optimization
- the swap body container transportation optimization problem (LHT network for parcel mail) [16]
- the reassignment of mail sources to sorting centers

This paper is also concerned with projects on the optimization of such subnetworks. We will describe the approaches and the results of three recent projects in more detail. We have selected these projects from different subsystems of the overall network, and we cover both the strategic and the tactical phases. The projects are in the areas of

- collection of mail from mailboxes (vehicle routing)
- replanning of delivery station locations (facility location combined with vehicle routing)
- reduction of deadheading in the last mile (sequential facility location and vehicle routing approach)

Each of the projects covers system analysis, modeling, development of optimization algorithms, and implementation of a software prototype (decision support system).

2 Optimization of Mailbox Collection Tours

In the collection part of a postal logistics network, mail is collected from different types of locations, e.g., mailboxes, business customers, and retail stores. The different types of pick-up points can be distinguished by characteristics, such as their service time windows, the collection frequency, or collected mail volume. In this section we focus on the collection of letter mail from mailboxes. Even though mail volume is generally small at mailboxes, mail often needs to be collected multiple times a day, as mailboxes may fill up quickly. This especially holds if they are located in busy spots, e.g., train stations. In Germany the collection times are noted on each mailbox individually, and mail may not be collected from a mailbox before this specific time. For this reason, service time windows in collection tours are determined by the earliest arrival time, which is the collection time displayed on a mailbox. The latest arrival time at a mailbox is not restricted as long as mail is not *overflowing* out of the mailbox.

Multiple studies concerning the optimization of mailbox collection tours have been published, e.g., by Laporte et al. [11] for Canada Post Corporation in Montreal, by Mechti et al. [12] for the French postal service, and by Tarantilis et al. [19] for some known benchmarks from the literature. None of them consider restricted vehicle capacities, as mail volume on a tour is small in general. However, the latest arrival time at a depot or the maximal duration of tours prevent the introduced models from visiting all mailboxes with a single tour.

2.1 Modeling and Solving

The optimization of mailbox collection tours may be modeled as a vehicle routing problem (VRP) [20, 21]. As mentioned above, for the optimization of mailbox collection tours at Deutsche Post DHL, time windows need to be considered. Also, as we are taking multiple sorting centers into consideration, we may change the allocation of mailboxes. This decision is implicitly made by assigning a mailbox to a tour, as each tour has a predefined start depot and end depot. Unlike other known models, in our application, sorting capacities at mail sorting centers are restricted. Yet, as collection tours may arrive until a defined final sorting time (cut-off time) has been reached, the restricted capacity cannot be modeled as a single resource. Much more, a certain arrival rate of mail at the sorting center must be met in order to guarantee completion of sorting before cut-off. For model complexity reasons, this arrival rate may be discretized into a set of points in time within the sorting time window of a depot. Then, the mail volume arriving at the depot with a tour later than a certain time is restricted. Figure 2 shows an example of discretization of sorting capacities for a generic sorting center, where $B^g(\tau)$ denotes the letter mail quantity that can be processed at depot g later than time τ.

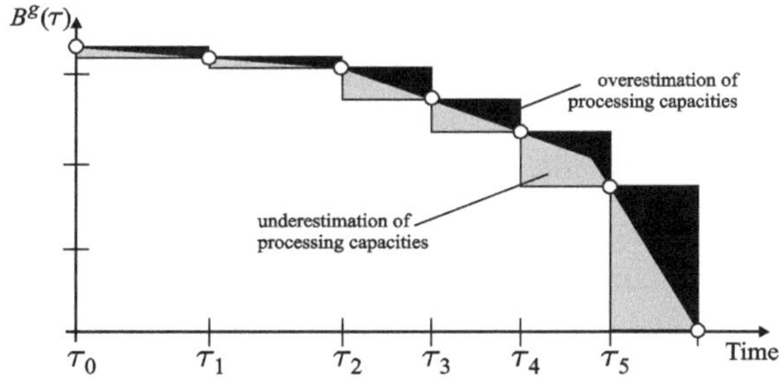

Fig. 2 Example of a discretization of a time-varying processing capacity function

The problem was modeled as a multi-depot vehicle routing problem, where the sorting capacities of the depots (sorting centers), which may vary over time, are taken into account as restricted inter-tour resources [6], i.e., resources that are restricted not to one tour but for all tours at the same time. The mathematical model and its parameters, decision variables, restrictions, and the objective function are described now.

Sets:

- C set of customer nodes
- G set of depot nodes
- K set of vehicles or tours
- V^k set of feasible nodes for tour $k \in K$
- A^k set of feasible arcs for tour $k \in K$
- L set of points in time

Parameters:

- $o(k)$ start node/origin of tour k
- $d(k)$ end node/destination of tour k
- a_i lower bound for a resource at node i
- b_i upper bound for a resource at node i
- t^k_{ij} resource demand on direct link from node i to j for tour $k \in K$
- $B^g(\tau_\ell)$ mail that can be processed at depot g later than time ℓ
- n_g number of tours that end at depot g
- $k(g,h)$ the hth vehicle ending at depot g

The objective (1) of the model minimizes the total costs accumulated along all tours. Constraints (2) ensure that each customer $i \in C$ is visited by one and only one tour. Each tour $k \in K$ contains a unique start depot $o(k)$ and also a unique end depot $d(k)$ (see constraints (3)). The flow conservation of a tour $k \in K$ at node $i \in V^k$ is represented by constraints (4). With the binary routing variables x^k_{ij} (5), constraints

Decision variables and resource vectors:

- $x^k_{i,j}$ binary variable indicating that arc (i, j) is visited by tour k
- $T^{k,cost}_{d(k)}$ accumulated costs of tour k at its destination depot $d(k)$
- $T^{k,load}_i$ accumulated mail pick-up volume at node i of tour k

$T_i^{k,time}$ accumulated travel and waiting time at node i of tour k

$S_{\ell,h}^{g}$ partial sum of all loads arriving at depot g later than time ℓ for the first $1, 2, \ldots, h$ tours

$f_{ij}(T_i^k)$ $\max\{a_i, T_i^k + t_{ij}^k\}$

(6) and (7) simply state that the paths $P = P(x^k)$ on each tour $k \in K$ have to form resource-feasible paths. Inequalities (8) guarantee that the sequence of partial sums is non-decreasing. Constraints (9) model the interdependency between arrival time and collected load of a tour, and the corresponding partial sum. If tour k with depot destination d_k arrives after τ_ℓ ($T_{d(k)}^{k,time} > \tau_\ell$), then the hth partial sum $S_{\ell,h}^{g}$ must exceed $S_{\ell,h-1}^{g}$ by the delivered volume $T_{d(k)}^{k,load}$ at depot d_k. If τ_ℓ ($T_{d(k)}^{k,time} \le \tau_\ell$) holds, then constraints (8) and (9) allows the setting of $S_{\ell,h}^{g} = S_{\ell,h-1}^{g}$. Through constraints (10), processing capacities after time ℓ are restricted. Note that constraints (8)–(10) are non-linear.

$$\min \sum_{k \in K} T_{d(k)}^{k,cost} \tag{1}$$

$$\text{s.t.} \sum_{k \in K} \sum_{j:(i,j) \in A^k} x_{ij}^k = 1 \quad \forall i \in C \tag{2}$$

$$\sum_{j:(o(k),j) \in A^k} x_{o(k),j}^k = \sum_{i:(i,d(k)) \in A^k} x_{i,d(k)}^k = 1 \quad \forall k \in K \tag{3}$$

$$\sum_{j:(i,j) \in A^k} x_{ij}^k - \sum_{j:(j,i) \in A^k} x_{ji}^k = 0 \quad \forall k \in K, i \in V^k \tag{4}$$

$$x_{ij}^k \in \{0, 1\} \quad \forall k \in K, (i, j) \in A^k \tag{5}$$

$$T_i^k \in [a_i, b_i] \quad \forall k \in K, i \in V^k \tag{6}$$

$$x_{ij}^k (f_{ij}(T_i^k) - T_j^k) \le 0 \quad \forall k \in K, (i, j) \in A^k. \tag{7}$$

$$S_{\ell,h-1}^{g} \le S_{\ell,h}^{g} \quad \forall g \in G, \ell \in L, h \in \{2, \ldots, n_g\} \tag{8}$$

$$(T_{d(k)}^{k,time} - \tau_\ell)(S_{\ell,h-1}^{g} + T_{d(k)}^{k,load} - S_{\ell,h}^{g}) \le 0 \quad \forall g \in G, \ell \in L, h \in \{2, \ldots, n_g\}, \; k = k(g, h) \tag{9}$$

$$0 \le S_{\ell,h}^{g} \le B^g(\tau_\ell) \quad \forall g \in G, \ell \in L, h \in \{2, \ldots, n_g\} \tag{10}$$

The model and its solution are based on the unified modeling and solution framework by Irnich [10]. Tours are connected to a giant tour, where resources, such as time and load, are restricted at each node. The consumption of resources along arcs or tour segments is calculated through so-called resource extension functions (REFs) (see [9]). To find a good feasible solution quickly for a problem instance, customer nodes are iteratively inserted into dummy tours and re-inserted into tours through a variable neighborhood descent (VND) algorithm. Also, iterations of destroy moves followed by the VND are used to diversify solutions and to explore a bigger part of the solution space. Details of the solution approach are described in [6, 10].

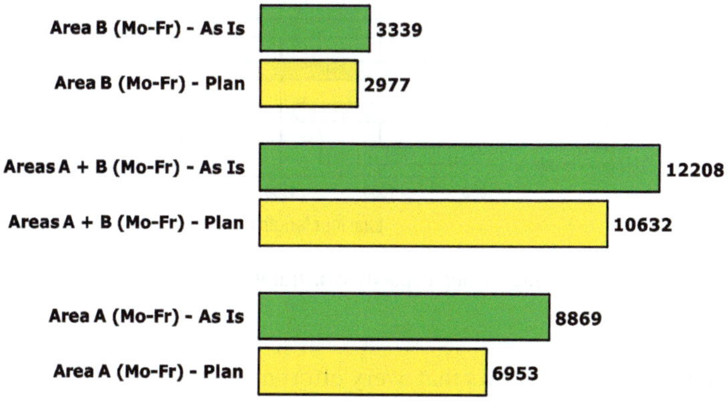

Fig. 3 Comparison of total tour length (in minutes) of today's tours and an optimized tour plan for two neighboring pilot regions (denoted as areas A and B)

2.2 Implementation

At Deutsche Post DHL, most mailbox collection tours start and end at a depot. Here, they pick up the keys for the mailboxes as well as a scanner, which they need to document the mailboxes visited along a collection tour. After visiting all mailboxes, the tour returns to the SC, bringing in the collected mail, the keys, and the scanner.

For the optimization of the collection tours, first the current tour plan is analyzed. The arrival rate of mail volume with the current tours arriving at an SC is set as the sorting center's capacity. Mail brought in by new tours must not exceed this arrival rate, i.e., with an optimized tour plan, more volume should arrive earlier than today at each SC.

For analysis purposes, we used real data of two neighboring regions representing urban and rural areas. Figure 3 shows the total time of today's tour plan and the result of our optimization for the pilot regions. The tours contain more than 2,000 stops, as each mailbox may be emptied multiple times per day. Through our optimization, more than 20 % of total tour time is saved in *Area A* (mostly urban areas) and about 10 % in *Area B* (mostly rural areas). The results show that this approach is suitable for urban areas as well as for rural areas. We assumed that, a simultaneous optimization of neighboring areas could better exploit the optimization potentials. When solving a multi-depot VRP with time windows (MDVRPTW) with two depots instead of two separate single depot vehicle routing problems with time windows, 13 % of total tour time may be saved in *Areas A + B*. Despite the promising result, the absolute decrease of total time is less than the sum for *Area A* and *Area B*. The reason for this result is the heuristic approach, the large number of stops for *Areas A + B* and therefore the growing complexity. The influence of time-varying sorting capacities on the total tour length is shown in Fig. 4. Even though the optimized tour plan (*MDVRPTW with capacity constraints*) is about 5 % better than today's tour plan (*Today*), tour length can be significantly shortened when relaxing or removing the

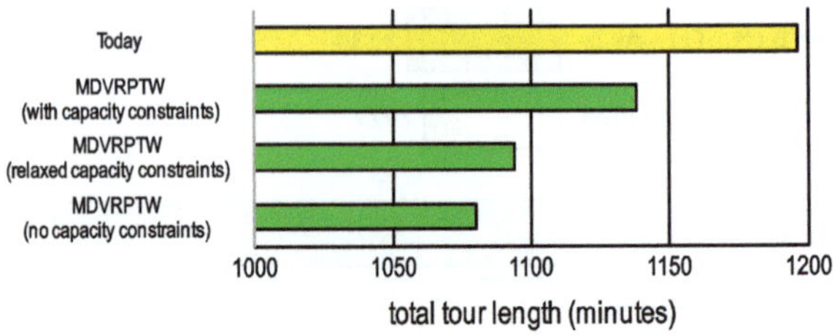

Fig. 4 Influence of time-varying sorting capacity constraints on total tour length

capacity constraints. This shows that every effort needs to be made to model sorting capacities as accurately as possible.

Thanks to the results of successful pilots with the implemented prototype, the model introduced above is today part of the software used for the operational planning of mailbox collection tours at Deutsche Post DHL. Since September 2008, Deutsche Post DHL has realized a significant cost reduction by optimizing the mailbox collection tours using this software.

3 Optimization of Delivery Stations

We consider now the subnetwork *distribution and delivery of the last mile*. Our goal is the optimization of the delivery processes within this subnetwork by strategic network design decisions. We introduced this network briefly in Sect. 1.1. The relevant objects are sorting centers and the delivery stations for letter mail. Delivery stations are mini-hubs where the final preparation (final sorting) of mail for the delivery takes place (performed by the postman) and where the postman starts the delivery tours covering his DD. Figure 5 shows a simplified schematic picture of the subnetwork *distribution and delivery of the last mile*.

In order to better understand the transportation link in stage 2 of the network, we consider Fig. 6. In reality, the postman moves from the DS to the first customer of

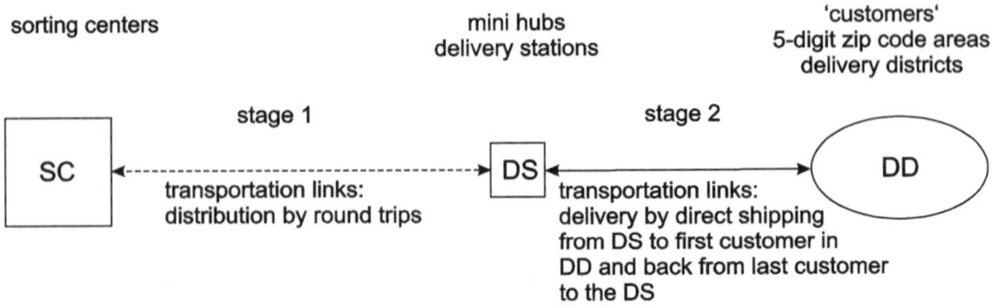

Fig. 5 Schematic view of the subnetwork *distribution and delivery of the last mile*

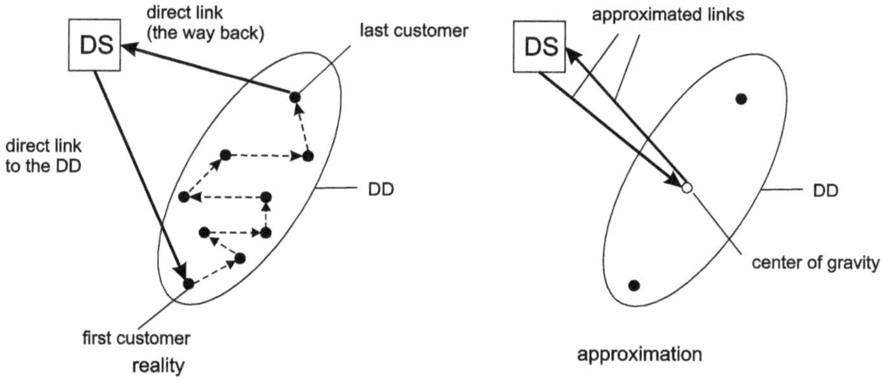

Fig. 6 Raw approximation of a delivery district by its center of gravity

the DD and after visiting all customers of the DD, the postman returns from the last customer to the DS. Of course, the DS to first customer, last customer to DS links are not productive from the standpoint of the core delivery process. They take time and therefore the length of these unproductive links should be minimized. In order to simplify the model, we concentrate the postman tours within the delivery districts by using the center of gravity of the DD instead of considering a first and a last customer of the postman tour. Then, we get an approximated distance: *DS to center of gravity of DD*. The transportation links in stage 1 (see Fig. 5) are much more complicated, because distribution from the SC to the assigned set of delivery stations is organized by round trips. We will consider the type of round trips used in stage 1 later in detail. After describing the network structure and the transportation processes, we have to define the optimization problem. Of course, one may think about different, more -or less- complex and complicated problems. We focus here on the following scenario:

- The sorting centers for letter mail belonging to a well defined geographic area are given (i.e. locations and capacities are given). The well defined geographic area is the area allocated to the considered SC (see Sect. 1.1).
- The allocated area considered above is composed of a set of 5-digit zip code areas and is, at the same time, cut into a set of delivery districts. These delivery districts and the routes used by the postman to visit them are assumed to be given.
- There is a set of potential DS locations consisting of existing delivery stations and new locations. Each location belonging to the set of potential DS locations is called a *candidate location*.

The problem is to select DS locations from the set of candidate locations such that service quality requirements (restrictions) are fulfilled and an overall cost function becomes minimal. We will call this problem in the following *selection of optimal DS locations*.

3.1 Mathematical Formulation and Solution Approach

The solution approach to the problem *selection of optimal DS locations* is model-based. The mathematical model and its parameters, decision variables, restrictions, and the objective function will be described now.

Notations:

J set of zip-code areas j
SI set of indices of potential delivery locations DL
SH set of indices of sorting centers SC
SS set of indices of other relevant locations SO
$SI = \{1, \dots, n_I\}$ n_I number of potential delivery locations
$SH = \{n_I + 1, \dots, n_I + n_H\}$ n_H number of SCs
$SS = \{n_I + n_H + 1, \dots, n_I + n_H + n_S\}$ n_S number of SOs

A potential delivery location i, where $i \in I$, is either a DS (letter mail network LN) or a delivery base (DB) (parcel mail network PN). Therefore, we get: $SI = SI^{\text{LN}} \cup SI^{\text{PN}}$ and $SI^{\text{SYN}} = SI^{\text{LN}} \cap SI^{\text{PN}}$. SI^{LN} and SI^{PN} denote the index sets of the potential delivery locations for the letter mail or the parcel mail network. SI^{SYN} contains those potential delivery locations which can be used for both networks. Thus, the model takes into account the synergies of both the letter mail and the parcel mail networks.

Parameters:

f_i fixed costs for opening delivery location i
f_i^o annual fixed costs for operating delivery location i
$\widetilde{y}_i$ 1 if DL i is open in the initial situation, otherwise 0

We distinguish between opening and annual operating fixed costs. Both costs occur if a potential DL is selected and does not exist in the current configuration. Otherwise only the annual operating fixed costs occur.

Decision variables:

$x_{ij} \in \{0, 1\}$ for $i \in SI$ and $j \in J$
$y_j \in \{0, 1\}$ for $i \in SI$

$x_{ij} = 1$ holds if zip-code area j is assigned to the potential DL i, $x_{ij} = 0$ otherwise. $y_i = 1$ holds if the potential delivery location i is selected ('open'), $y_i = 0$ otherwise.

Restrictions:

$$\sum_{i \in SI} x_{ij} = 1 \; \forall \; j \in J \tag{11}$$

Each customer's demand (zip-code area $j \in J$) is completely fulfilled by delivery locations SI. Together with $x_{ij} \in \{0, 1\}$, Eq. (11) means *single sourcing*. Each zip-code area is completely assigned to exactly one DL.

$$\sum_{i \in SI^{\mathrm{LN}}} x_{ij} = 0 \; \forall \; j \in J^{\mathrm{PN}} \tag{12}$$

$$\sum_{i \in SI^{\mathrm{PN}}} x_{ij} = 0 \; \forall \; j \in J^{\mathrm{LN}} \tag{13}$$

The set J of zip-code areas contains, for example 5-digit zip-code areas j. A zip-code area of this kind is covered by delivery districts for letter or for parcel mail separately and also by districts which are designed for combined letter and parcel mail delivery. J^{PN} denotes the set of delivery districts for parcel mail delivery only and J^{LN} the set for letter mail delivery only. Then, (12) means that a delivery district for parcel mail only cannot be allocated to a delivery location for letter mail only and vice versa (13). In addition, we have a capacitated problem:

$$\delta_{ij} x_{ij} \leq d_i^{max} \; \forall \; i \in \; SI \text{ and } j \in J \tag{14}$$

This means that the demand δ_{ij} of zip-code area j for a sorting area within the DL $i \in SI$ must be smaller than the capacity d_i^{max} of the DL. If the potential DL i is not open, $y_i = 0$, then it is not allowed to assign zip-code areas to this not-selected delivery facility.

$$x_{ij} \leq y_i \; \forall \; i \in SI \text{ and } j \in J \tag{15}$$

Objective function:

$$\min z = \sum_{i \in SI} \sum_{j \in J} c_{ij} x_{ij} + \sum_{i \in SI} \left(f_i (1 - \tilde{y}_i) y_i + f_i^o y_i \right), \tag{16}$$

whereby c_{ij} represents allocation costs if the zip-code area j is assigned to the DL $i \in SI$ (costs for the links from delivery location i to the center of gravity of a DD belonging to j, aggregated over all delivery districts belonging to j).

The term $\left(f_i (1 - \tilde{y}_i) y_i + f_i^o y_i \right)$ describes annual fixed costs related to DL i and the fixed costs for opening the location i. In the case where delivery location i already exists ($\tilde{y}_i = 1$), only annual operating fixed costs f_i^o apply.

The mathematical model can be characterized as a two-stage facility location problem. Also, it considers the synergies of both the letter and the parcel mail networks, which makes it interesting on the one hand but also very complex on the other. Problem instances become too big for computing exact solutions using the existing solvers and hardware. The approach is adapted to a replanning problem in contrast to a complete new network design task. This means that some of the existing DS/DB locations will remain stable while others are questionable. Also, there are existing round trips with related costs in the first distribution stage. These round trips should be taken into account either by the model or by the solution approach, which extends the problem in the direction of a location routing problem [13].

Concluding, we decided to develop a (meta-)heuristic approach, which was designed and implemented by Hermanns [7]. Next, we explain this iterative 3-phase heuristic approach.

- Start with the existing solution (distribution/delivery network). The algorithm starts with the existing delivery locations and routes, checks feasibility, and computes the related costs.
- Each iteration consists of 3 phases in the sequence:

1. location phase
2. allocation phase
3. routing phase

Location Phase

Selection of locations (from the set of potential delivery locations) which appear *attractive*, to be used as DS/DB-facilities. This selection is independent of allocation or routing decisions.

Allocation Phase

The locations chosen in phase 1 are now given. Customers (meaning delivery districts) are allocated to these locations. If there is no feasible allocation (e.g. because the capacities of the facilities selected in phase 1 are too small in order to satisfy the customer demand) the solution determined in phase 1 cannot be accepted. If a termination criterion is not fulfilled after diversification/intensivation steps, phase 1 must be repeated. In order to check the feasibility of the decisions made in phases 1 and 2 and to compute the related cost, the mathematical model described above is used.

Routing Phase

From the existing solution we know an existing configuration of locations (and related facilities), an allocation, and an existing set of routes (start = iteration 0). The same applies after each iteration i. After phases 1 and 2 of iteration i+1, we know the changes in locations and allocations compared to iteration i. Therefore, the routes belonging to iteration i have now to be modified such that the location and allocation decisions of iteration i+1 are taken into account.

After phases 1 to 3, a feasible solution has been constructed. If this solution is better than the best known solution up to iteration i+1, it becomes the new best known solution. Otherwise, the algorithm either stops or continues with a diversification step.

In the following we present a modified approach, which can be characterized thus:

- The location phase (phase 1) controls the algorithm. We use different operators e.g. the 'add' and 'drop' operators, which characterize the neighborhoods for local search. Also, we introduce a diversification strategy in phase 1.

- After selecting an operation in phase 1 (e.g. adding a closed candidate DL), we solve the resulting allocation problem in phase 2. This can be done by using a model-based approach applying a commercial MIP solver or by specialized algorithms [7].
- Now, within this add/drop-loop (phase 1) we know the related optimal allocation decisions and are therefore able to move to phase 3 in order to determine the best related routing decisions. This is done by operations which modify the existing routes by a route in the neighborhood.

Figure 10 shows the flow of the used heuristic. In the prototype tool for planning DS locations (TOPAS), the allocation problem in phase 2 is solved by a simple heuristic (nearest location) in the uncapacitated case and by a knapsack algorithm in the capacitated case. The overall algorithm is implemented as a tabu search and as a simulated annealing metaheuristic as well [7]. The most interesting component of the algorithm is the modification of the existing routes (known from the previous iteration) taking into account the new location/allocation decisions. We will illustrate the TOPAS approach to this problem using an example. First, we show the complexity of routes in this application area.

Example illustration

Let us assume that a route t^0 contains several locations. We denote by

- $H = \{SC1\}$ the set of sorting centers,
- $I = \{DS1, DS2, DS3, DS4\}$ the set of potential delivery stations, and
- $S = \{SO1, SO2\}$ set of other relevant (for the first distribution stage) locations.

In Fig. 7 the initial route $t^0 = (SC1, SO1, DS1, DS2, SO2, DS3)$ is illustrated. Now, we consider an add move in phase 1, which adds $DS4$. In phase 2, customer 2 becomes allocated to $DS4$. Then, Fig. 8 shows a new tour t^1, which is generated by inserting $DS4$ between $DS1$ and $DS2$ and by deleting the direct link from $DS1$ to $DS2$.

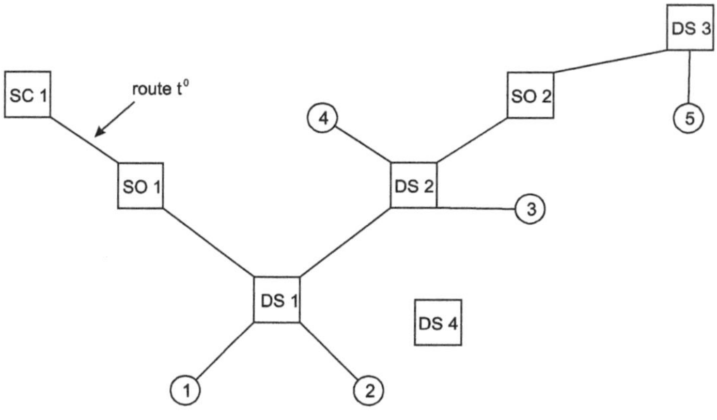

Fig. 7 An initial route t^0

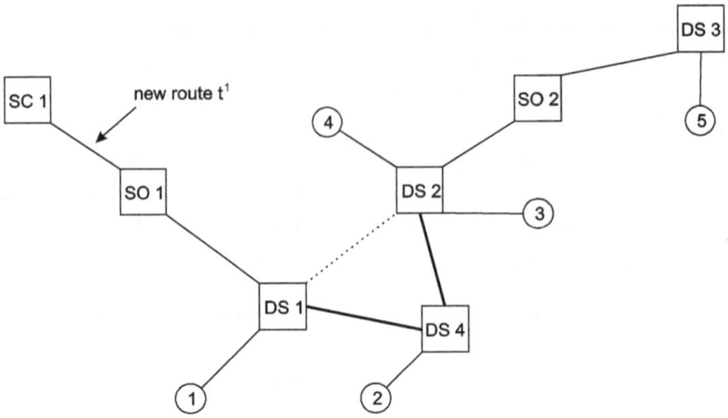

Fig. 8 New route $t^1 = (SC1, SO1, DS1, DS4, DS2, SO2, DS3)$ after an add move in phase 1, re-allocation in phase 2, and an insertion step in phase 3

Finally, in order to illustrate that different operations in phase 1 and different types of routes require different algorithmic solutions, we consider a second example, where the operator in phase 1 is the drop move and the considered tours are so-called central tours from the sorting center $SC1$ to delivery locations $DS1$, $DS2$, and $DS3$. The delivery stations have customers j_1, j_2, and j_3 allocated to them, which represent zip-code areas or regions composed of zip code areas. After dropping $DS1$, the tour $t2^0$ becomes shorter $t2^1$. In iteration 2, because of the additional dropping of $DS2$, all three customers j_1, j_2, and j_3 must be allocated to $DS3$. The tours $t1^2$ and $t2^2$ are now identical (see Fig. 9). All algorithms for the different categories are described in detail in [7].

The software prototype TOPAS was first applied for the optimization of the delivery stations in the year 2008. Since 2009, Deutsche Post DHL has achieved extensive economies by replanning the delivery station locations through using the prototype tool TOPAS. The implementation of TOPAS at Deutsche Post DHL, algorithms, numerical test and results, and economic results are described in [7] on pp. 139–160.

4 Reducing Deadheading on Postman Tours

In this section we consider now a part of the last mile of the letter mail network in Germany. Postmen start their workday at a DS with administrative tasks and the sorting of all/some letters according to the route they take when actually delivering mail. Since the DS is not necessarily located inside a postman's district, the postman tour may start with deadheading (from the DS to the first delivery point, see Fig. 11). It typically also ends with deadheading (from the last delivery point to the DS). In order to increase productivity, a reduction of these unproductive parts of the tour is reasonable. One way would be to set up many delivery stations close to or within each DD, but this would lead to increasing distribution costs (of stage 2, see Sect. 1.1) and

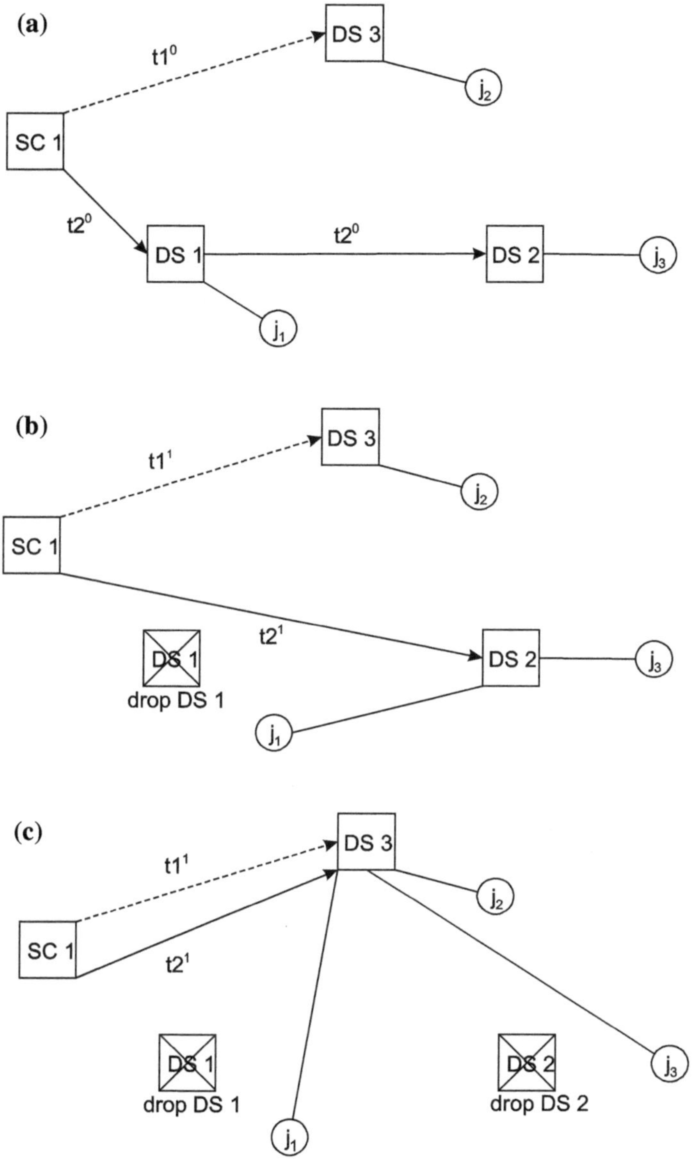

Fig. 9 Drop operations to different routes. **a** Two initial routes $t1^0$ and $t2^0$. **b** First iteration, one possible result after drop DS 1. **c** Second iteration, one possible result after drop DS 2

overhead/operating costs. Therefore, we are addressing the issue of how unproductive parts of postman tours can be reduced without introducing more delivery stations.

The solution to this problem is in the reorganization of the processes: The sorting and delivery of mail should be done by different employees. In this way, the

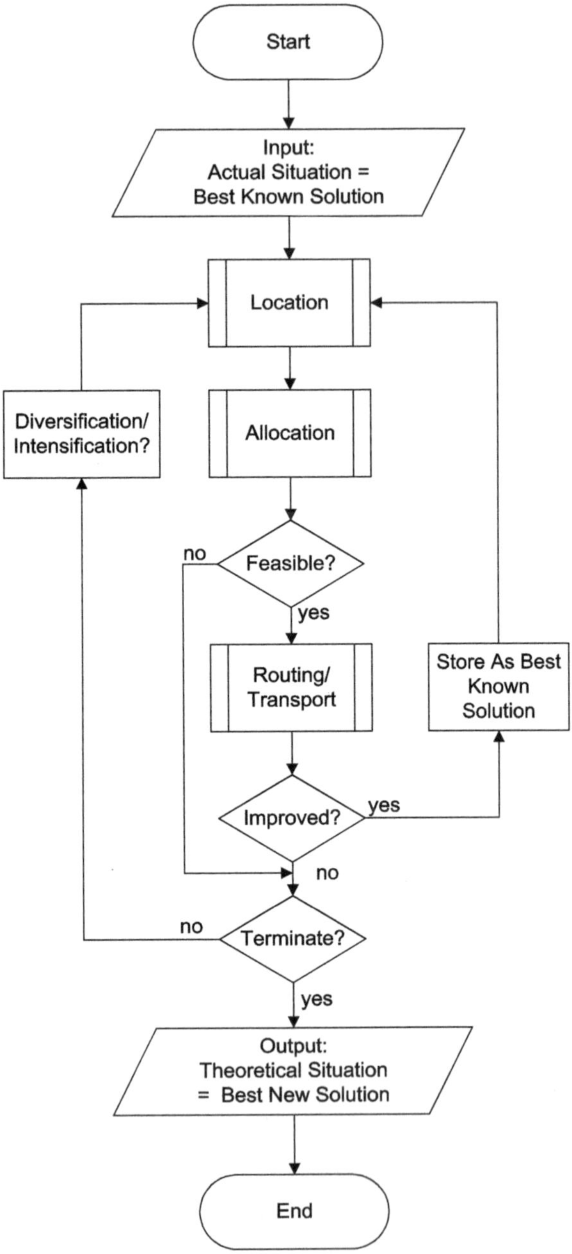

Fig. 10 Flow chart of TOPAS solution approach

preparation and some kind of sorting can be performed in the DS by specialized employees. Then, sorted mail is packed into boxes and shipped to transfer points by car. There, postmen take over the boxes and start the delivery within the delivery districts. An advantage of this solution is that transfer points have low operating

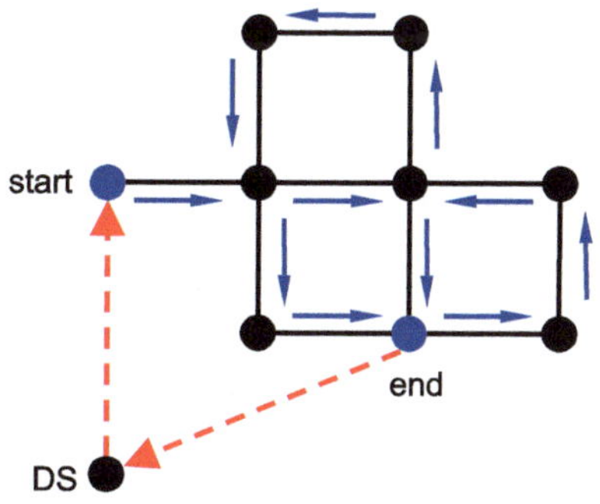

Fig. 11 Deadheading on a postman tour

costs and can be placed close to the delivery districts. Thus, postmen do not start their tours with deadheading, or at least they start with less deadheading. Finally, the task is to find the optimal number and location of transfer points, such that the sum of deadheading, transportation, and operational costs for transfer points is minimized (and the overall costs are lower than the costs for deadheading from the DS).

4.1 Mathematical Formulation

In order to model the problem, we assume that the set I of delivery districts and the set J of potential transfer points are known. Furthermore, operational costs (fix and variable costs) for transfer points, deadheading costs between each transfer point (TP) and DD, and shipment costs are known. Our approach for modeling and solving this problem is based on location routing theory [15] but it proceeds in two sequential steps: the first step is the determination of the number and location of transfer points, and the second step is the optimization of mail transportation costs from the DS to the transfer points.

The objective of the first step is to minimize the sum of deadheading and operational costs, whereby the following restrictions must hold: Each DD must be uniquely assigned to a TP. Further, because of employment laws, at least two postmen must start at the same TP. Additionally, due to space shortage at transfer points (these are, e.g., garages, car ports) a maximum number of postmen can work at the same TP. This problem is a capacitated warehouse location problem with single-sourcing constraints derived from [1]. Delivery districts relate to customers with demand 1 and transfer points to warehouses, where the capacity is defined by the maximum number of postmen at a TP. The mathematical formulation and its parameters, decision variables, restrictions, and the objective function will be described now.

Notations:

I set of delivery districts
J set of potential transfer points

The set of delivery districts is known in advance and the division of the delivery area is not a part of this problem. Further, the discrete set of potential transfer points is a subset of nodes within the street network of the delivery area.

Parameters:

f_j fixed costs of TP j
v_j variable costs of TP j
c_{ij} costs for deadheading between TP j and DD i
a_j minimum number of delivery districts assigned to an open TP j
b_j maximum number of delivery districts assigned to an open TP j

Fixed costs represent rent or leasing costs of a TP. Variable costs represent expenses per postman at a TP. Costs for deadheading are calculated as the minimum of the shortest paths from the TP j to each node of the DD i within the street network of the delivery area. The minimum of the shortest paths is multiplied by two (includes the way to and from the DD) and (time) travel costs. As mentioned above, the parameters a_j and b_j restrict the number of assigned delivery districts (postmen) to a TP j.

Decision variables:

$y_j \in \{0, 1\}$ binary variable indicating whether TP j is open or closed
$x_{ij} \in \{0, 1\}$ binary variable indicating whether DD i is assigned to TP j

$x_{ij} = 1$ holds if DD i is assigned to the potential TP j, $x_{ij} = 0$ otherwise. $y_j = 1$ holds if the potential TP j is selected ('open'), $y_j = 0$ otherwise.

$$\min \sum_{j \in J} f_j y_j + \sum_{j \in J} \sum_{i \in I} v_j x_{ij} + \sum_{j \in J} \sum_{i \in I} c_{ij} x_{ij} \tag{17}$$

$$\text{s.t.} \sum_{j \in J} x_{ij} = 1 \quad \forall \ i \in I \tag{18}$$

$$x_{ij} \leq y_j \quad \forall \ i \in I, j \in J \tag{19}$$

$$a_j y_j \leq \sum_{i \in I} x_{ij} \quad \forall \ j \in J \tag{20}$$

$$\sum_{i \in I} x_{ij} \leq b_j y_j \quad \forall \ j \in J \tag{21}$$

$$x_{ij} \in \{0, 1\} \quad \forall \ i \in I, j \in J \tag{22}$$

$$y_j \in \{0, 1\} \quad \forall \ j \in J. \tag{23}$$

The objective function (17) is minimizing the sum of fixed, variable, and deadheading costs. Constraints (18) represent the single-sourcing constraints, which means that each DD must be assigned to exactly one TP. On the other hand, constraints (19) restrict the assignment of a DD to a TP if and only if the TP is selected.

A minimum number of delivery districts must be assigned to a selected TP (20), and the number of assigned delivery districts cannot exceed a given upper bound (21). Constraints (22) and (23) represent the binary requirement of the decision variables. This problem was solved with the commercial solver MOPS [18].

Once the number and location of transfer points are determined, the task of the second step is to ship the sorted mail to the transfer points. Service quality aspects force postmen to start their delivery at the first delivery point at approximately 8.00 a.m. and the sorting of mail to be finished by 6.30 a.m. in the DS. So the transportation of sorted mail to transfer points can be performed only during the time window from 6.30 to 8.00 a.m. This problem relates to a vehicle routing problem with time windows (VRPTW) (see [2]) and its solution approach is based on the unified modeling and solution framework by Irnich [10].

4.2 Implementation

For application scope, both described problems and used solving methods were implemented in a prototype. For a given set of delivery districts and potential transfer points, the optimal number and location of transfer points and route plans from the DS to the transfer points can be optimized. Further, for a period of time (from Monday to Saturday) the overall cost (sum of deadheading, operational, and shipment costs) can be compared to the cost of deadheading from the DS of the former situation. Figure 12 shows an example scenario for reducing deadheading with the prototype. The prototype window is composed of four areas and will be described in the following. For simplicity, we call the situation before installing transfer points *current state* and the situation after installing transfer points *new state*.

The first area (Fig. 12a) contains information about each DD. In the order of the columns we have the identifier, the mode of transport (by bicycle or on foot) in the current state, the mode of transport in the new state, the deadheading in meters of the current state, the deadheading in meters of the new state, the deadheading in minutes of the current state, and finally the deadheading in minutes of the new state. The last two rows contain the sum and average of deadheading in meters and minutes of both states.

We retrieve the optimization data from the second area (Fig. 12b). In the first column we have the identifier of the selected transfer points. The following columns contain for each chosen TP its fixed costs, variable costs, minimum and maximum number of possible assignments, and the number of assigned delivery districts and their identifiers.

From the third area of the prototype (Fig. 12c) we gain cost information on the current and new states for each scenario calculation. In the order of the columns we have the identifier, the number of chosen transfer points, the sum of operating costs per week, the sum of deadheading costs per week, the number of needed tours for mail shipment to the transfer points, shipment costs per week, overall costs per week of the new state, and finally overall deadheading costs of the current state.

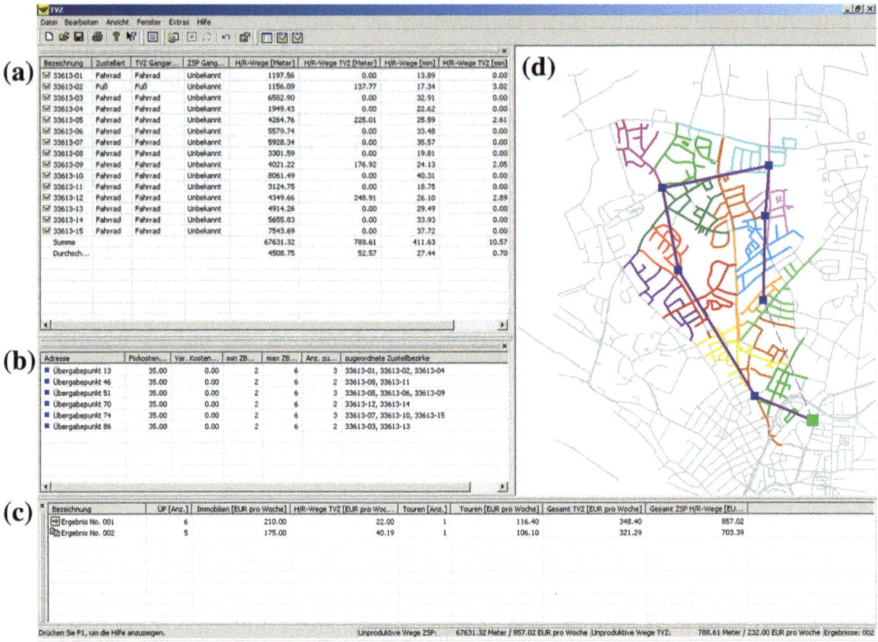

Fig. 12 Example scenario of deadheading reduction

The last area (Fig. 12d) is a visualization of the street network and delivery districts. Further, the selected transfer points are represented by small quadratic nodes and the DS is represented by a big quadratic node. The sequence of transportation tours from the DS is represented by connection lines, whereby the dashed line represents the return to the DS after the last visited TP on the tour. Moreover, the prototype can be used for scenario analysis. First, if there is no information available about potential transfer points, a set of potential transfer points is generated automatically on a grid pattern and shifted to the nearest node on the street network. The refinement of the grid pattern can be changed by the user if necessary. After the solution of this automatically generated scenario, local planners can retrieve information on where and how many TPs should be located. As mentioned above, transfer points are, e.g., car ports, whose availability at the locations suggested by the prototype has to be checked. In the majority this is not the case and, therefore, available transfer points close to the suggested ones must be found. After available potential transfer points are found and imported into the prototype, a new scenario can be computed and compared to the previous one(s). Furthermore, the prototype allows the user to manually insert potential transfer points and to shift them by their coordinates. An additional option window allows the user to fix transfer points, i.e. they have to be selected by optimization, or to change the default values of a_j and b_j individually. All described functionalities enable the generation and comparison of several different scenarios. This way, the prototype is a decision support system for retrieving information on where to search for suitable transfer points, where and how many to locate, and how to assign delivery districts to them. In addition, the prototype can be used to verify whether an already existing set of transfer points and its assignment of delivery districts is still optimal or at least a good solution. This is necessary in periodic cycles, because new potential transfer points could be available, or actual costs changes could occur.

Since 2006, the prototype has been applied to more than 4,000 delivery districts and Deutsche Post DHL has saved significant expenses by reducing deadheading on their postman tours.

5 Conclusion

The distribution networks in the postal logistics area are very complex. Therefore, decomposition into planning phases and subnetworks are necessary in order to optimize the distribution networks. In detail we have described three successful projects which have been executed by the Deutsche Post Chair of Optimization of Distribution Networks at RWTH Aachen University and by Deutsche Post DHL. These projects are examples of successful OR approaches in practice, consisting of problem analysis, algebraic optimization model development, solution of the optimization problems (using standard software tools or metaheuristics as well), and development of software prototypes. This paper does not only contain well known OR models and algorithms, but it also contributes to the development of methods and algorithms. In particular, it contains an approach to the multi-depot vehicle routing problem with restricted inter-tour resources, a location-routing approach for replanning problems by tabu search and a capacitated warehouse location-routing problem.

Each of the three projects was also successful from an economic point of view. Extensive costs savings were achieved by the replanning of the subnetworks described above using the three prototypes. At the same time the high service level was maintained.

References

1. Balinski ML (1965) Integer programming: methods, uses, computation. Manag Sci 12:253–313
2. Bräysy O, Gendreau M (2005) Vehicle routing with time windows, part I: route construction and local search algorithms. Transp Sci 39:104–118
3. Büdenbender K, Grünert T, Sebastian HJ (2000) A hybrid tabu search branch and bound algorithm for the direct flight network design problem. Transp Sci 34:364–380
4. Engelhard G, Grünert T, Sebastian HJ (1998) Thäringen M, Katz M, Kuchem R: Und ab geht die Post—Transportplanung für den Brieftransport der Deutschen Post AG. OR News
5. Grünert T, Sebastian HJ (2000) Planning models for long-haul operations of postal and express shipment companies. Eur J Oper Res 122:289–309

6. Hempsch C, Irnich S (2008) Vehicle-routing problems with inter-tour resource constraints. In: Golden BL, Raghavan S, Wasil EA (eds) The vehicle routing problem: latest advances and new challenges. Springer, New York, pp 421–444
7. Hermanns C (2009) Planung und Optimierung von Auslieferungsstandorten in komplexen Distributionsnetzwerken. Mainz Verlag, Aachen
8. Irnich S (2002) Netzwerk-Design für zweistufige Transportsysteme und ein Branch-and-Price-Verfahren für das gemischte Direkt- und Hinflugproblem. Dissertation, RWTH Aachen University, http://darwin.bth.rwth-aachen.de/opus3/volltexte/2002/300/pdf/Irnich_Stefan.pdf. Accessed 08 Nov 2013
9. Irnich S (2007) Resource extension functions: properties, inversion, and generalization to segments. OR Spectr 30:113–148
10. Irnich S (2008) A unified modeling and solution framework for vehicle routing and local search-based metaheuristics. Informs J Comput 20:270–287
11. Laporte G, Chapleau S, Landry PE, Mercure H (1989) An algorithm for the design of mailbox collection routes in Urban areas. Transp Res B-Methodol 23:271–280
12. Mechti R, Poujade S, Roucairol C, Lemarie B (1999) Global and local moves in tabu search: a real-life mail collection application. In: Voß S, Martello S, Osman IH, Roucairol C (eds) Meta-Heuristics: advances and trends in local search paradigms for optimization. Kluwer Academic, Boston, pp 155–174
13. Nagy G, Salhi S (2007) Location-routing: issues, models and methods. Eur J Oper Res 177:649–672
14. Pajunas A, Matto EJ, Trick M, Zuluaga LF (2007) Optimizing highway transportation at the United States postal service. Interfaces 37:515–525
15. Perl J, Daskin MS (1985) A warehouse location-routing problem. Transp Res B 19:381–396
16. Sebastian HJ (2012) Optimization in postal logistics. Presented at the INFORMS conference on applying science to the art of business, Miami, FL, http://meetings2.informs.org/Practice06/FINAL%20progPractice%20Conference%202006.pdf. Accessed 08 Nov 2013
17. Sebastian HJ (2012) Optimization approaches in the strategic and tactical planning of networks. In: Dolk D, Granat J (eds) Modeling for decision support in network-based services. Lecture Notes in Business Information Processing. Springer, Berlin Heidelberg, pp 36–61
18. Suhl UH (1994) MOPS—mathematical optimization system. Eur J Oper Res 72:312–322
19. Tarantilis CD, Kiranoudis CT, Markatos NC (2002) Use of the BATA algorithm and MIS to solve the mail carrier problem. Appl Math Model 26:481–500
20. Toth P, Vigo D (2002) An overview of vehicle routing problems. In: Toth P, Vigo D (eds) The vehicle routing problem. Siam, Philadelphia, pp 1–23
21. Toth P, Vigo D (2002) Branch-and-bound algorithms for the capacitated VRP. In: Toth P, Vigo D (eds) The vehicle routing problem. Siam, Philadelphia, pp 29–51

Optimizing Long-Haul Transportation Considering Alternative Transportation Routes Within a Parcel Distribution Network

Matthias Meisen

Abstract Due to increasing costs, intense competition and increasingly demanding customer expectations regarding delivery lead times parcel delivery services need to continuously improve their logistics networks and processes. This paper analyses the potential to improve both costs and delivery lead times by introducing alternative, direct transportation routes. Here, direct transports avoid hubs within the distribution network and, therefore, reduce transportation distance and time as well as sorting time. To examine the optimization potential of these new transportation routes, we consider a large-scale distribution network, where distinguishable items have to be transported from sources through certain hubs to predetermined sinks. We introduce a service network design model, which minimizes total long-haul transportation costs within the delivery network while meeting predetermined service levels. The model identifies the optimal transportation plan and decides if a service is offered at a specific time period or not.

1 Introduction

The German parcel market has been facing some major structural changes during the past years. While well-known direct commerce companies went out of business new e-commerce companies, online auction sites and others have been helping to increase the number of parcel shipments in Germany by 66 % from 1995 to 2007 [19]. At the same time parcel services are exposed to the following challenges:

- The adoption of toll for trucks on the Autobahn, increasing diesel prices, increasing wages and labour cost as well as new working time regulations in transportation regarding driving time and rest periods have been leading to **increasing costs**.
- The expansion of parcel networks by most parcel services operating in Germany have been leading to an **intense competition with a high pricing pressure**. While the German consumer price index increased in total by almost 25 % during the

M. Meisen (✉)
Deutsche Post DHL, Charles-de-Gaulle-Str. 20, 53113 Bonn, Germany
e-mail: Matthias.Meisen@deutschepost.de

H.-J. Sebastian et al. (eds.), *Quantitative Approaches in Logistics and Supply Chain Management*, Lecture Notes in Logistics,
DOI 10.1007/978-3-319-12856-6_6

last 15 years and by 7 % from 2005 to 2009, the price index for parcel services decreased by more than 3 % in the same time period [20].

- The increasing costs and price pressure are accomplished by **demanding customer expectations**. This especially becomes noticeable in higher expectations regarding the quality of services such as pickup and delivery as well as the claim for value-added guaranteed services and the desire for shorter delivery lead times and a more reliable compliance of the D + 1-delivery.[1]

All of the above mentioned challenges force parcel services to continuously optimize their logistics networks. The presented paper is motivated by a project at Deutsche Post DHL where ways to reduce delivery lead times in the existing parcel network were examined. One possible solution is the introduction of direct transportation allowing selected transports to skip hubs and, thus, save both transportation and sorting time. To identify the potential of direct transportation within the parcel network of Deutsche Post DHL a mixed-integer linear program is introduced. A solution to this model offers an optimal transportation plan with respect to cost while meeting a predetermined service level, i.e., percentage of D + 1-deliveries. To integrate service level constraints into the model, it is necessary to include not only classical flow conservation and capacity constraints to the model but also the notion of time. Yet, this leads to a vast number of additional variables and constraints. However, clever limitations of the solution space allow for the model to be implemented in AIMMS [1] and solved with CPLEX [9].

2 The German Parcel Delivery Network of Deutsche Post DHL

With the arrival of the "Postreform" II in July 1995, the Deutsche Bundespost was converted into a private enterprise with the aim of enabling greater efficiency, better prices and better services [21]. The focus of this transformation was based on a parcel network consisting of 33 hubs, which still are the core of DP DHL's parcel distribution network. Besides these hubs, the network consists of approximately 200 delivery depots (German: Zustellbasis (ZB)), 3,000 delivery stations (German: Zustellstützpunkt (ZSP)), 14.000 local affiliates, 2,500 Packstations and 1,000 Paketboxes. A simplified representation of the parcel distribution network of Deutsche Post DHL is shown in Fig. 1. One characteristic of the network is the partition into collection, long-haul transportation and distribution. The regional pick-up at local affiliates, Packstations, Paketboxes or at customers is being held in the collection. After being collected all parcels are shipped to one of the 33 outbound hubs (German: Paketzentrum Abgang (PZA)), where all items undergo an automated outbound sorting in which the parcels are sorted by its particular hub destination. During and after the outbound sorting, the nationwide transport between the hubs, which is called long-haul, takes place. In total there are $33 \times 32 = 1{,}056$ long-haul transportation links that are served daily by Deutsche Post DHL. A second sorting takes place in the inbound

[1] D + 1 means that a parcel is delivered one day after being handed over to a parcel service.

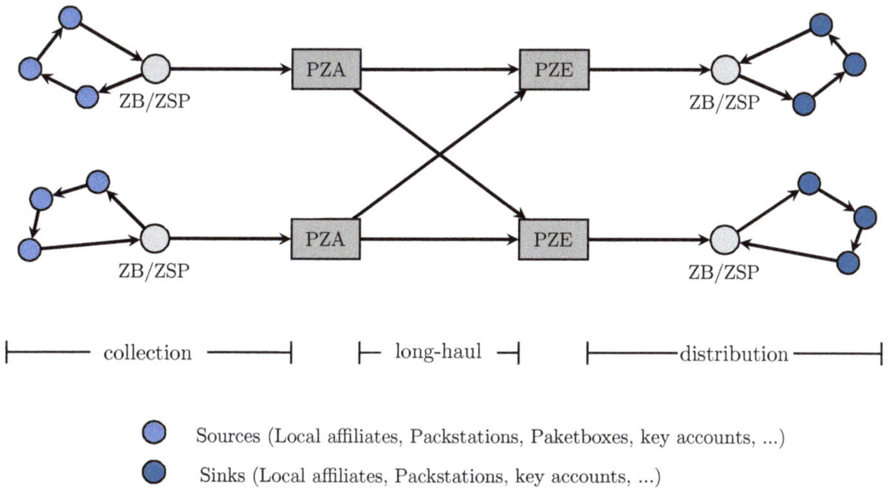

Fig. 1 A simplified representation of the parcel delivery network in Germany

hub (German: Paketzentrum Eingang (PZE)) where the parcels are sorted according to their respective delivery depot or delivery station. After being transported to a delivery depot or delivery station all parcels are delivered to their recipients. In collection as well as in distribution the allocation of a delivery depot/delivery station to a certain hub is unique.

2.1 Possible Extensions of the Network

In order to fulfill the customers' desire for shorter delivery lead times, parcel services like Deutsche Post DHL are continuously looking for ways to optimize their parcel networks. One possible way to reduce lead times as well as to increase service levels might be the extension of existing distribution networks by direct transportation which leads to alternative transportation routes. Direct transportation services skip hubs and, therefore, reduce transportation distance as well as transportation time which might lead to decreasing production costs and an increasing service level. In order to allow direct transportations the shipped parcels have to be prepared for skipping hubs by a more accurate sorting. Looking at the structural organization of the distribution network of Deutsche Post DHL (see Fig. 1), delivery depots and delivery stations are the first possibility within the parcel network where an additional sorting process could take place. The basic prerequisite for a "pre-sorting" is that the delivery depot or delivery station is equipped with appropriate sorting equipment. Due to the small physical sizes of delivery stations, such modifications can only be realized in delivery depots. In the case of an adequate sorting in a delivery depot, the sorted parcels can be shipped directly from a delivery depot to another delivery depot (see Fig. 2a) or from a delivery depot to an inbound hub (see Fig. 2b). Furthermore, a possible refinement of the outbound sorting in the outbound hub enables direct shipments from the outbound hub to a particular delivery depot (see Fig. 2c). Due to further sorting processes that are needed in the destination, direct shipments to delivery stations are not realizable. According to our explanations, there are three possible direct shipments in the parcel network that cause the following **alternative transportation routes**:

- Outbound delivery depot (ZB) → Delivery depot (ZB)
- Outbound delivery depot (ZB) → Inbound hub (PZE) → Delivery depot (ZB)
- Outbound delivery depot (ZB) → Outbound hub (PZA) → Delivery depot (ZB)

Extending the parcel network by these long-haul routes lead to the network illustrated in Fig. 2d. The described alternative transportation routes do have the following main advantages:

1. **A faster nationwide transport**: Due to direct transports, detours are avoided and the holding time is reduced. This leads to a faster nationwide transport and shorter lead times which might increase the D + 1-ratio.
2. **A relief of sorting capacities**: Due to skipping hubs, sorting capacities in various hubs are relieved. This is especially advantageous in sorting centers where the capacity is a bottleneck. Thus, direct shipments lead to an increase of the D + 1-ratio by relieving the sorting capacities.

However, direct shipments do not only cause advantages. One main disadvantage is that the shipped quantity outgoing from a delivery depot is usually small.

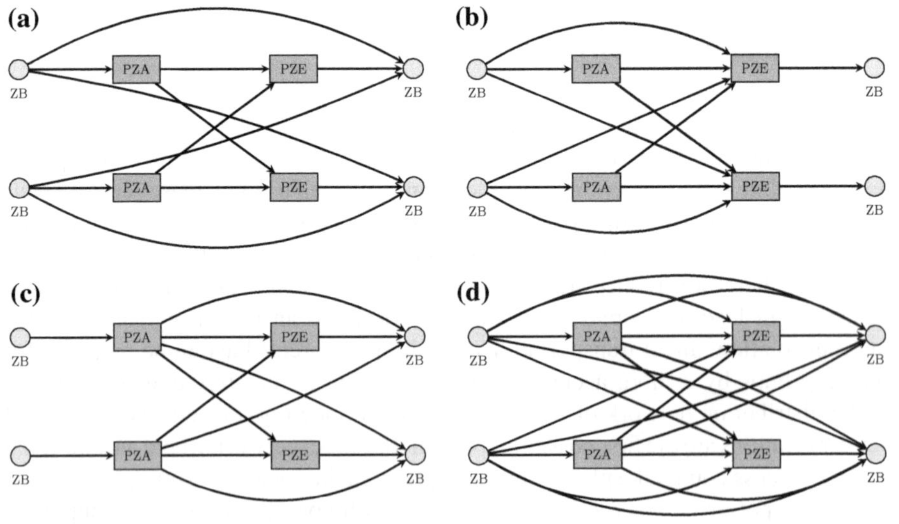

Fig. 2 Direct transportation links within a distribution network. **a** Direct transportation from ZB to ZB. **b** Direct transportation from ZB to PZE. **c** Direct transportation from PZA to ZB. **d** The extended parcel network

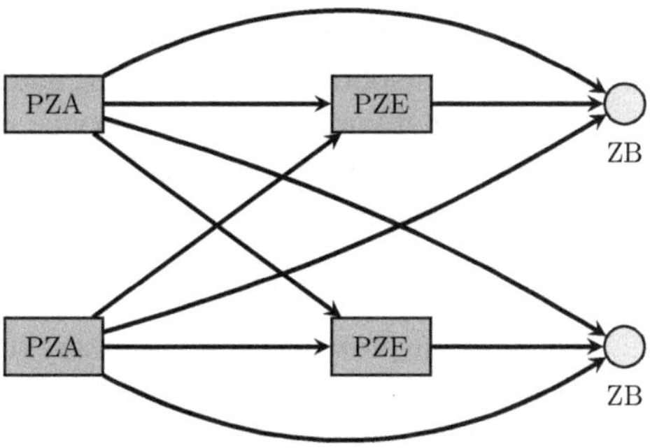

Fig. 3 A network extension through direct transportation links

Accordingly, the capacity of direct shipments starting from a delivery depot would not fully be used. Thus, cost savings resulting from shorter transportation distances are counterbalanced by unused effects of consolidation. Moreover, delivery depots have to be equipped with particular sorters in order to sort parcels. Such a retooling is attended by costs and the economical benefit has to be proven for every delivery depot independently. Finally, a pre-sorting in a delivery depot leads to an additional sorting process. Although some parcels would possibly be sorted for direct shipment, the remaining parcels are subjected to an additional sorting process without provoking any extra benefits. As a result, a pre-sorting in delivery depots might lead to higher sorting costs and longer lead times. The mentioned disadvantages show that direct transports departing from a delivery depot are not reasonable. Therefore, from now on we will disregard direct transports starting from a delivery depot. Instead, we will focus on direct shipments from outbound hubs to various delivery depots as shown in Fig. 3.

3 Network Design

Due to the mentioned challenges within the German parcel market, parcel services are forced to optimize their logistics networks in order to stay competitive. Therefore, an appropriate network design has an essential influence on the success of a parcel service, since it does not only influence the type and quality of an offered service but also determines the resulting costs [15]. In dependence of the importance and the time horizon of the decisions that have to be met, Crainic and Laporte [13] distinguish between strategic (long-term), tactical (medium-term) and operational (short-term) planning. As can be seen from the last section, we focus on tactical planning, which deals with the design of service networks. Services can be transports or a repositioning of vehicles.

Service network design models have been widely and intensively discussed in the literature. Besides basic model formulations as can be seen in Kim [16] or Irnich [15], service network design models can be found in the literature for various transportation modes. There exist service network design models for railway transportation (see Cordeau et al. [8]), maritime transportation (see Christiansen et al. [6, 7]), long-haul transportation (see Crainic [10] or Crainic and Laporte [13]) as well as for multimodal transportation (see Crainic and Kim [12]). Despite the many publications focussing on network design and service network design, in the literature there are only a few publications dealing with the application of these models for planning parcel and postal networks. Exceptions are the publications by Armacost et al. [2], Barnhart and Schneur [4], Barnhart et al. [3], Cheung et al. [5], Kim et al. [17], Kuby and Gray [18] or Irnich [15].

Besides transportation costs, in many real-life transportation problems the factor time plays a crucial role. This always holds, when shipped items need to arrive before a particular time at a certain place. For considering arrival and departure times in service network design models, the model formulation needs to be extended by the dimension time transforming static models into dynamic models. An essential methodical tool for modelling deterministic dynamic network design problems are discrete time-space-networks [15]. In these models, the planning period is discretized in an adequate number of periods and a vertex is created for all physical locations in each time period [14]. Thus, a time-space-network arises by duplicating the vertex set $\mathscr{V}$ according to the periods of time. This means that a vertex v_t corresponds to a location v within the time period t. An arc between the vertices v_t and $w_{t'}$ represents either

- a service between various sites v and w that departs at time t from site v and arrives at time t' or
- a sorting/handling of shipments at location $v = w$ during the time from t to t'.

Time-space-networks normally do have a large number of vertices and edges, even though the number of physical locations might be quite moderate [15]. Both, the size of the network as well as additional time constraints complicate the solving of these models [11]. Thus, often heuristics or metaheuristics are used to solve deterministic dynamic service network design models [14].

Until now we have only considered a physical parcel network, consisting of vertices as locations and edges as transportation links. In order to consider arrival and departure times of services, the network is extended by the dimension time leading to a time-space-network shown in Fig. 4. The solid arcs in the graph describe transports

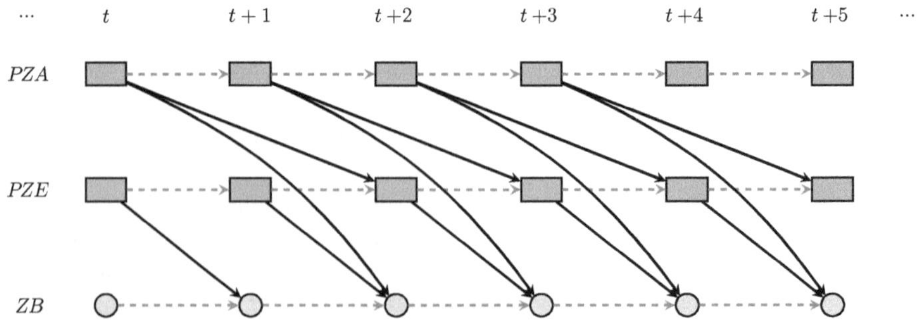

Fig. 4 The extended network as Time-Space-Network

that depart from location i at time t and arrive at location j some periods later. Thus, the arrival time depends on the departure time and the duration t_{ij} of the offered transportation service between i and j. The dotted lines describe procedures (e.g. sorting processes) taking place at a particular location. Although the time-space-network in Fig. 4 only includes one outbound hub (PZA), one inbound hub (PZE) and a single delivery depot (ZB), it clearly demonstrates the size and complexity of time-space-networks considering real-life sized parcel networks. Furthermore, the modelling of time-space networks gets even more complicated by not only considering transports but also heterogeneous goods shipped on the particular transportation links.

4 A Service Network Design Model for Optimizing Long-Haul Transportation

We now introduce a mixed integer program, which minimizes the total long-haul transportation cost within the in Sect. 2.1 described large-scale distribution network. The model solution identifies the optimal transportation plan and decides in dependency of a given service level, whether a service is offered at a specific time period or not. Thus, we need flow conservation and capacity constraints as well as time constraints.

4.1 Model Formulation

In order to formulate a service network design model we used the following assumptions:

- The total shipment is available in the hubs at time period $t = 0$. To ensure a realistic view on the sorting process, we assumed that all items are sorted in the outbound hub at the latest possible time during the sorting process. Under this assumption we can calculate the number of sorted parcels in an outbound hub in advance. This number is given in parameter s_u^t.
- We also assumed that additional costs of a more accurate sorting in the PZA will be balanced through savings that result in the PZE due to a reduced number of items that have to be sorted. Thus, we only consider transportation costs.
- Moreover, we assumed that the sorting capacity over the considered time period is sufficient enough to sort all items.
- While sorting the parcels, we do not know the number of items with a given destination that will be sorted in a certain time period. Thus, we assumed that for each commodity k at every time period t the same percentage ratio φ_i^k is given in hub i. The ratio φ_i^k can be calculated by dividing the shipped items of commodity k at hub i by the total number of shipped items of site i. These priori calculated ratios change when direct transportations take place which skip the inbound sorting.

Since these changes cannot be described linear, we assumed these ratios to be fixed.

For modelling the problem we will consider the large-scale network as a directed graph $\mathcal{N} = (\mathcal{V}, \mathcal{A})$. The vertices $\mathcal{V}$ of the network $\mathcal{N}$ will be classified into the sets $\mathcal{U}$ (set of all outbound hubs), $\mathcal{W}$ (set of all inbound hubs) and $\mathcal{D}$ (set of all delivery depots and delivery stations). Thus, $u \in \mathcal{U}$, $w \in \mathcal{W}$ and $d \in \mathcal{D}$ represent a vertex of the respective set. Furthermore, the set $\mathcal{A}$ describes the set of all arcs (respectively all transportation links) within the distribution network. The set $\mathcal{K}$ represents the several commodities that are shipped through the network. Thereby, the parcels can be distinguished by its respective destination within the network $\mathcal{N}$. Last but not least, the set $\mathcal{T}$ includes all time periods that are considered in the model. In the following we will list all sets and parameters.

Sets:

$\mathcal{A} = \{1, ..., A\}$: Set of all transportation routes (arcs)
$\mathcal{U} = \{1, ..., U\}$: Set of all outbound hubs
$\mathcal{W} = \{1, ..., W\}$: Set of all inbound hubs
$\mathcal{D} = \{1, ..., D\}$: Set of all delivery depots and delivery stations
$\mathcal{K} = \{1, ..., K\}$: Set of all commodities (parcels with destination k)
$\mathcal{T} = \{0, ..., T\}$: Set of all time periods in the time-space network
$\mathcal{F} = \{1, ..., F\}$: Set of all means of transport
$V(i)$: Set of all predecessors of vertex $i \in \mathcal{V}$
$N(i)$: Set of all successors of vertex $i \in \mathcal{V}$

Parameters:

a_u^k: Quantity of items k to be shipped from outbound hub u
b_d^k: Received items k at destination d
μ_w^t: Sorting capacity at inbound hub w
κ^f: Capacity given by means of transport f
c_{ij}^f: Cost of a service on link (i, j) using means of transport f
t_{ij}: Transportation time on link (i, j)
ς: Given service level (D + 1-ratio)
φ_i^k: Ratio of the items with destination k at site i
t_{ES_i}: Time of sorting end at site i
s_u^t: Bulk of sorted parcels in outbound hub u at time t
t_{D+1}: Parcels, that arrive at a delivery depot or delivery station before t_{D+1} will be delivered within D + 1

The variables of our model are $x_{ij}^{t,k}$, $y_{ij}^{t,f}$, q_i^t and s_w^t. Variable $x_{ij}^{t,k}$ describes the flow of items k between two sites i and j at time t. $y_{ij}^{t,f}$ describes a binary variable, which decides whether a service f is arranged between i and j at time t. The variable q_i^t models the quantity of the arriving shipments in location i at time t, whereas s_w^t describes the amount of parcels sorted at time t in the PZE. With these sets, parameters, variables and assumptions we can now formulate a service network design model for optimizing the long-haul transportation:

$$\min \quad \sum_{(i,j)\in\mathcal{A}} \sum_{t\in\mathcal{T}} \sum_{f\in\mathcal{F}} y_{ij}^{t,f} \cdot c_{ij}^{f} \tag{1}$$

$$\text{s.t.} \quad \sum_{i\in N(u)} \sum_{t\in\mathcal{T}} x_{ui}^{t,k} = a_u^k \quad \forall u \in \mathcal{U}, \quad k \in \mathcal{K} \tag{2}$$

$$\sum_{i\in V(d)} \sum_{t\in\mathcal{T}} x_{id}^{t,k} = b_d^k \quad \forall d \in \mathcal{D}, \ k \in \mathcal{K} \tag{3}$$

$$\sum_{i\in V(j)} \sum_{t\in\mathcal{T}} x_{ij}^{t,k} = \sum_{l\in N(j)} \sum_{t\in\mathcal{T}} x_{jl}^{t,k} \quad \forall j \in \mathcal{W}, \quad k \in \mathcal{K} \tag{4}$$

$$\sum_{k\in\mathcal{K}} x_{ij}^{t,k} \leq \sum_{f\in\mathcal{F}} y_{ij}^{f,t} \cdot \kappa^f \quad \forall (i,j) \in \mathcal{A}, \quad t \in \mathcal{T} \tag{5}$$

$$q_j^t = \sum_{i\in V(j)} \sum_{k\in\mathcal{K}} x_{ij}^{t-t_{ij},k} \quad \forall j \in \{\mathcal{W} \cup \mathcal{D}\}, \quad t \in \mathcal{T} \tag{6}$$

$$s_w^t \leq \mu_w^t \quad \forall w \in \mathcal{W}, \quad t \in \mathcal{T} \tag{7}$$

$$\sum_{t\in\mathcal{T}} s_w^t = \sum_{t\in\mathcal{T}} q_w^t \quad \forall w \in \mathcal{W} \tag{8}$$

$$\sum_{\tau=0}^{t} s_w^\tau \leq \sum_{\tau=0}^{t} q_w^\tau \quad \forall w \in \mathcal{W}, \quad t \in \mathcal{T} \tag{9}$$

$$\sum_{\tau=t}^{t_{ES_w}} q_w^\tau \leq \sum_{\tau=t}^{t_{ES_w}} \mu_w^\tau \quad \forall w \in \mathcal{W}, \quad t \in \mathcal{T} \tag{10}$$

$$\sum_{j\in N(i)} x_{ij}^{t,k} \leq \sum_{\tau=0}^{t} s_i^\tau \cdot \varphi_i^k - \sum_{\tau=0}^{t-1} \sum_{j\in N(i)} x_{ij}^{\tau,k}$$
$$\forall i \in \{\mathcal{U} \cup \mathcal{W}\}, \quad t \in \mathcal{T}, \quad k \in \mathcal{K} \tag{11}$$

$$\sum_{l\in N(j)} x_{jl}^{t,k} \leq \sum_{\tau=0}^{t_{ES_w}} \sum_{i\in V(j)} x_{ij}^{\tau-t_{ij},k} - \sum_{\tau=0}^{t-1} \sum_{l\in N(j)} x_{jl}^{\tau,k}$$
$$\forall j \in \mathcal{W}, \ t \in \mathcal{T} \text{ mit } t > t_{ES_w}, \quad k \in \mathcal{K} \tag{12}$$

$$\sum_{d\in\mathcal{D}} \sum_{\tau=0}^{t_{D+1}} q_d^\tau \geq \varsigma \cdot \sum_{d\in\mathcal{D}} \sum_{k\in\mathcal{K}} b_d^k \tag{13}$$

$$x_{ij}^{t,k} \geq 0 \quad \forall (i,j) \in \mathcal{A}, \quad t \in \mathcal{T}, \quad k \in \mathcal{K} \tag{14}$$

$$y_{ij}^{t,f} \in \{0,1\} \quad \forall (i,j) \in \mathcal{A}, \quad t \in \mathcal{T}, \quad f \in \mathcal{F} \tag{15}$$

$$q_i^t \geq 0 \quad \forall i \in \{\mathcal{W} \cup \mathcal{D}\}, \quad t \in \mathcal{T} \tag{16}$$

$$s_w^t \geq 0 \quad \forall w \in \mathcal{W}, \quad t \in \mathcal{T} \tag{17}$$

The objective function of the model minimizes the sum of all transportation costs. Restrictions (2)–(4) form the flow conservation constraints. Thus, constraints (2) ensure that all parcels leave the respective outbound hub whereas constraints (3) make sure that all shipped items arrive at their designated destination. Equations (4) guarantee the flow conservation of shipped parcels in all inbound hubs. Thus, all parcels that are shipped to an inbound hub have to leave the hub after being sorted. Nevertheless, a flow is only possible if a certain transportation capacity is provided by a service which is modelled by the inequalities (5). Moreover, these constraints restrict the maximum flow on a transportation link by the provided transportation capacity. Restrictions (6) describe the arriving number of shipped items at the respective site which can be an inbound hub or a delivery station or a delivery depot. The sorting process in the inbound hub is modelled by restrictions (7)–(10). Thereby, the maximum number of sorted parcels per time period is restricted by the provided sorting capacity of a hub (see restrictions (7)). Furthermore, all parcels need to be sorted in the inbound hub (see constraints (8)) but the number of sorted parcels until a certain time period is limited by the sum of already arrived parcels (see restrictions (9)). To make sure that all parcels can be sorted in inbound hubs, inequalities (10) guarantee a sufficient remaining sorting capacity for sorting all parcels which have not been sorted or arrived yet. Moreover, restrictions (11) model the outgoing flow from a hub. This flow is restricted by the number of parcels that have been sorted but not shipped. The constraints (12) enable a balancing of shipped items after the end of the sorting process. This balance is necessary since the fixed assumed ratios φ_w^k may slightly change due to direct transportations. The predetermined service level is modelled by inequality (13). By changing ς, the given service level can be in- or decreased. In addition, constraints (14)–(17) describe the used variables. For the flows $x_{ij}^{t,k}$, the arriving parcels q_i^t as well as for the sorted parcels s_w^t non-negativity constraints hold (see constraints (14), (16) and (17)). Finally, (15) models a binary variable, which decides whether a service is arranged or not. As for all basic network flow models the supply needs to equal the demand for every commodity $k \in \mathcal{K}$. Thus,

$$\sum_{u \in \mathcal{U}} a_u^k = \sum_{d \in \mathcal{D}} b_d^k \quad \forall\, k \in \mathcal{K}$$

holds, which is checked in advance.

4.2 Adjustments to the Model and Reduction of Variables

The above-introduced mixed-integer problem is very large for realistic problem instances and a satisfactory solution is only realizable with an extensive computing time. As a realistic problem instance, we consider a network having 33 outbound hubs, approximately 200 delivery depots and about 245 delivery stations. In this large-scale distribution network exist about 8,000 transportation links at every time period and, thus, there are about 8,000 binary decision variables at every time t. The time is discretized into 29 time periods with a duration from 30 min each. In order to reduce the number of variables and constraints, we now present further adaptions to the model.

The first adaption is the limitation of direct transports to the 20 largest delivery depots, whereby $33 \times 180 = 5{,}940$ binary variables can be dropped at each time t. Moreover, all flow variables $x_{ij}^{t,k}$ contain the information, which commodity k flows on link (i, j). At an outbound hub, this distinction is only relevant for those parcels that can be shipped directly to a delivery depot. For all other items we only need the information to which inbound hub the parcels have to be shipped. Since all incoming items at the inbound hub are determined in the variable q_w^t, we can merge all commodities that need to be shipped exclusively over a particular hub. Merging these items lead to a total of 53 commodities (20 delivery depots, 33 inbound hubs) instead of 445 commodities that have to be considered at an outbound hub. Thus, merging diverse commodities causes a reduction of $33 \times 392 = 12{,}936$ flow variables at every time t in each outbound hub. Nevertheless, at the inbound hub you have to distinguish between all 445 commodities again. In addition, you can also a priori forbid the flow of shipments on certain links. Since there exists a unique allocation of a delivery depots and delivery stations to an inbound hub in the distribution, it is only allowed to ship parcels to a particular inbound hub whose destination is a delivery depot or a delivery station that is allocated to the particular hub. Therefore, we introduce set $Z(i) \; \forall \, i \in \{\mathscr{W} \cup \mathscr{D}\}$ which consists of all commodities $k \in \mathscr{K}$, that are allocated to a particular inbound hub or a particular delivery depot or delivery station.

For adjusting our model according to these considerations, we introduce the sets $\mathscr{D}_1$ and $\mathscr{D}_2$ as well as the sets $\mathscr{K}_1$, $\mathscr{K}_2$ and $\mathscr{K}_3$. The set $\mathscr{D}_1$ contains all delivery depots, to which a direct transportation can be arranged. All other delivery depots and delivery stations are combined in the set $\mathscr{D}_2$. Similarly, we introduce the sets $\mathscr{K}_1$, $\mathscr{K}_2$ and $\mathscr{K}_3$. The set $\mathscr{K}_1$ consists of all commodities whose destination are in set $\mathscr{D}_1$, the set $\mathscr{K}_2$ includes all commodities that are shipped to the sites $\mathscr{D}_2$ and $\mathscr{K}_3$ consists of all merged commodities, that are shipped to an inbound hub. Accordingly, d_1 represents a vertex from set $\mathscr{D}_1$, d_2 describes a vertex from set $\mathscr{D}_2$ and k_r represents a commodity out of set $\mathscr{K}_r \; \forall \, r \in \{1, 2, 3\}$. Simultaneously, we consider the allocation from commodities to a respective vertex of the network. Thus, only these commodities that are allocated to site (j) can flow on transportation link (i, j). Figure 5 illustrates the introduced sets and allocations.

Since the flow from a source to a sink can be either directly or over a particular inbound hub, flow conservation at the inbound hub holds automatically. Thus, the flow conservation constraints at the hub can be disregarded.

A further reduction of variables arises from limitating departure times of services in the model to real departure times of services. It is useless to offer outgoing services when there is almost no shipment available at a particular production site. Therefore, we assumed that outgoing services from a hub are scheduled between t_{BA_i} and t_{EA_i}. Since services are needed in order to have a flow, flow variables can also be limited by departure times of real services.

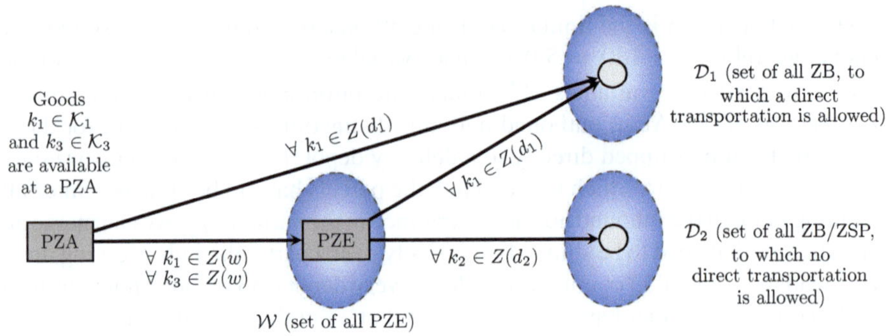

Fig. 5 Illustration of the introduced Sets

Limitations concerning the time are also reasonable for some constraints. Thus, all constraints regarding sorting processes can be adjusted to the period in which the sorting process takes place. This leads to the fact that possibly not all parcels can be sorted at an inbound hub and, thus, cannot be transported to a delivery depot or a delivery station so that the demand at these sites is not fulfilled. To avoid the violation of these flow constraints, we introduce a time period t_M where additional services can be operated to balance shipments. The time t_M is chosen in such a way that $t_M > t_{D+1}$ holds. Thus, a service can be arranged at time t_M, but shipped parcels cannot be delivered with D + 1. Moreover, the arriving time at delivery depots or delivery stations will not be considered. The only thing of interest is the fact that a balance of the items happens and that transportation costs for these services are included in the model. Moreover, since we do not consider any sorting costs, parcels leaving the PZA at time t_M do not need to be sorted in the model and, thus, the remaining sorting capacity does not need to suffice to sort all parcels. Therefore, both mentioned restrictions can be removed from the model formulation.

Due to our changes, the restrictions from our latter model can be adjusted in order to reduce the number of variables as well as the number of constraints. Since the adjusted model is very large and the individual restrictions are pretty similar to the already presented constraints we only want to give a few exemplary restrictions to show how the adjustments are being implemented. Therefore, we will describe restrictions dealing with outgoing flows at an outbound hub. In our model, each outbound hub is having a particular supply that has to be shipped away, either to an inbound hub or to a delivery depot. The Eqs. (18) and (19) guarantee that all items $k_1 \in \mathscr{K}_1$ and $k_3 \in \mathscr{K}_3$ will be shipped away from outbound hubs.

$$\sum_{i \in \{\mathscr{W} \cup \mathscr{D}_1\}} \sum_{t \in \mathscr{T}} x_{ui}^{t,k_1} = a_u^{k_1} \quad \forall\, u \in \mathcal{U}, \quad k_1 \in \mathscr{K}_1 \tag{18}$$

$$\sum_{w \in \mathscr{W}} \sum_{t \in \mathscr{T}} x_{uw}^{t,k_3} = a_u^{k_3} \quad \forall\, u \in \mathcal{U}, \quad k_3 \in \mathscr{K}_3 \tag{19}$$

Nevertheless, the outgoing flow needs to be restricted to the number of sorted parcels that have not been shipped yet. Therefore, constraints (20) and (21) are introduced.

$$\sum_{i \in N(u)} x_{ui}^{t,k_1} \leq \sum_{\tau=0}^{t} s_u^\tau \cdot \varphi_u^{k_1} - \sum_{\tau=0}^{t-1} \sum_{i \in N(u)} x_{ui}^{\tau,k_1}$$
$$\forall\, u \in \mathscr{U}, \quad k_1 \in \mathscr{K}_1, \quad t_{BA_u} \leq t \leq t_{EA_u} \tag{20}$$

$$\sum_{w \in N(u)} x_{uw}^{t,k_3} \leq \sum_{\tau=0}^{t} s_u^\tau \cdot \varphi_u^{k_3} - \sum_{\tau=0}^{t-1} \sum_{w \in N(u)} x_{uw}^{\tau,k_3}$$
$$\forall\, u \in \mathscr{U}, \quad k_3 \in \mathscr{K}_3, \quad t_{BA_u} \leq t \leq t_{EA_u} \tag{21}$$

These restrictions hold exactly in that time, where transports take place (respectively between t_{BA_i} and t_{EA_i}). Similar adjustments can be done to all constraints from the latter model with the aim of reducing the number of variables and constraints. In the end, the presented model without any direct transports consists of 263,406 constraints and 160,903 variables whereof 11,476 are binary. In the case of direct transports to the 20 biggest delivery depots the model includes 266,046 constraints and 166,183 variables whereof 16,756 are binary.

5 Results

After having implemented the adjusted model by using the modelling language AIMMS 3.10 [1], the problem is solved by using CPLEX 12.1 [9]. The computational time is limited to 10,800 s for every instance. Although the following results are based on a problem instance that is comparable from its structure and its size to the parcel distribution network of Deutsche Post DHL, the results are not showing real transportation costs from Deutsche Post DHL for the German parcel network.

5.1 The Baseline Scenario

In the baseline scenario we consider a parcel delivery network as described in Sect. 2, where a direct transportation from an outbound hub to delivery depots is not allowed. Thus, parcels have to be first shipped from an outbound hub to an inbound hub and then from the specific inbound hub to an assigned delivery depot or delivery station.

Figure 6 shows the daily transportation cost of parcel distribution starting from an outbound hub in dependence on a given D + 1-ratio. It can be seen that the transportation cost at the beginning increases only slowly with an increase of service level. Thus, there are only slight cost differences between a D + 1-ratio of about 80 % and

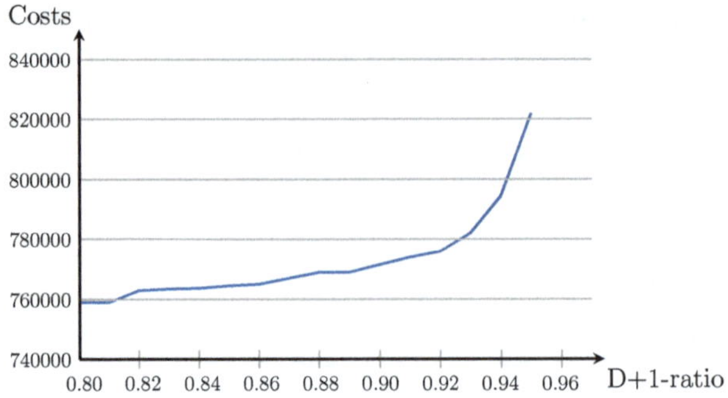

Fig. 6 Transportation costs in dependency of a given D + 1-ratio

a D + 1-ratio of about 92 %. This result can be explained by the fact that only a few more services with a high degree of capacity utilization have to be arranged between inbound sorting centers and delivery depots or delivery stations to increase the service level. This observation changes with an increasing service level. Thus, a higher number of services with a lower degree of capacity utilization are needed for a further uprating of the D + 1-ratio. In detail, the transportation cost rises by 1.6 % (3.4 %) in order to increase the D + 1-ratio from 93–94 % (94–95 %). At the same time Fig. 6 shows that a maximum D + 1-ratio from 95 % can be realized in the baseline scenario. Accordingly, 5 % of all shipped parcels in this scenario can not be delivered within the next day due to transportation distances between sorting centers as well as bottlenecks in the sorting process.

Moreover, Fig. 6 illustrates that transportation costs tend to € 750,000 by decreasing the service level. In the formulated model, every parcel needs to be transported from a source to a sink. Even if a D + 1-ratio of 0 % is pretended, transportation costs from outbound sorting centers to final destinations are arising. This transportation cost describes the most economical distribution of shipped parcels without any time constraints.

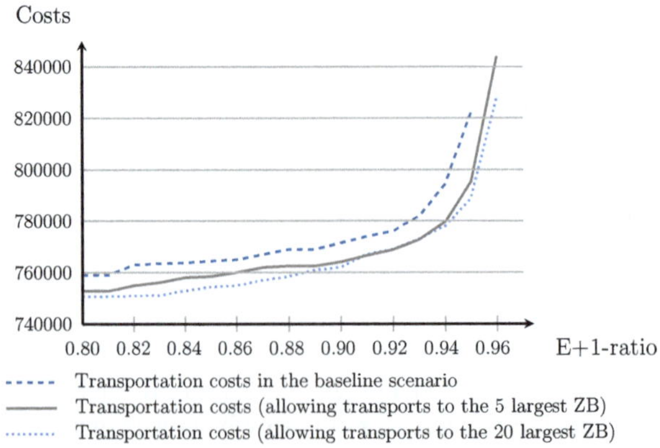

Fig. 7 Transportation costs with and without direct transportations

Table 1 Cost savings and service level improvements through direct transports in scenario 1

(a) Cost savings through direct transports	
D + 1-ratio in the baseline scenario (%)	Cost savings through direct transportations (%)
80	1.7
81	1.5
82	1.6
83	1.6
84	1.4
85	1.3
86	1.4
87	1.3
88	1.4
89	1.0
90	1.2
91	0.8
92	0.9
93	1.2
94	2.5
95	4.1
(b) Service level improvements	
D + 1-ratio in the baseline scenario (%)	Improvements of the D + 1-ratio in percentage points through direct transportations
90	2.63
91	2.48
92	2.05
93	1.50
94	1.14
93	0.83

5.2 Scenario 1: Direct Transportation

After describing a distribution network with no alternative transportation routes allowed in the baseline scenario, direct transportations from outbound hubs to delivery depots are permitted in scenario 1. The number of delivery depots which can be approached directly can be changed within the scenario. Thus, only delivery depots which are not integrated to a hub are considered, so that direct transportations have a real impact on the reduction of transportation distances. Figure 7 shows a comparison between transportation costs in the baseline scenario and transportation costs permitting direct transportations to the 20, respectively the 5 largest delivery depots. It can be seen that transportation costs in scenario 1 are lower than in the baseline scenario.

A closer examination of the calculated results lead to the in Table 1a listed potential of cost savings caused by direct transportations to the 20 largest delivery depots.

According to the results, the transportation costs can be reduced by 1.5 % on average due to direct transportations. Moreover, Fig. 7 shows the transportation costs allowing only direct transportations to the 5 largest delivery depots. The illustration clarifies that a significant part of cost savings already arise by allowing direct transportations to only a few delivery depots. Thus, transportation costs can be reduced by 1.1 % on average when allowing direct transportations to the 5 largest delivery depots. These computational results show that the transportation costs decrease the more direct transportations are allowed but the marginal benefit reduces by a decreasing delivery depot size.

Alternatively, a service level increase in the parcel distribution can be contemplated without generating additional transportation costs. Table 1b shows service level improvements plotted against the D + 1-ratio of the baseline scenario. Although the potential of optimization declines by an increase of the service level, the D + 1-ratio of the baseline scenario in the range from 90–95 % can be improved on average by 1.77 % points due to direct transportations. The main reasons for the existing potential of optimization are a release of sorting capacity helping to avoid or rather reduce bottlenecks in the hubs as well as the reduction of transportation distances whereby parcels arrive earlier at their final destination.

5.3 Scenario 2: Impact of an Increasing Shipping Volume

Between 1995 and 2007 the total volume of parcel shipments in Germany increased by 66 % [19]. Assuming a continuing trend the quantity of parcel shipments will increase even more during the next years. Thus, scenario 2 elaborates the impact of an increasing number of shipments to transportation costs. For the calculation, we assume a 10 % increase in total shipments. At first glance, the cost trends shown

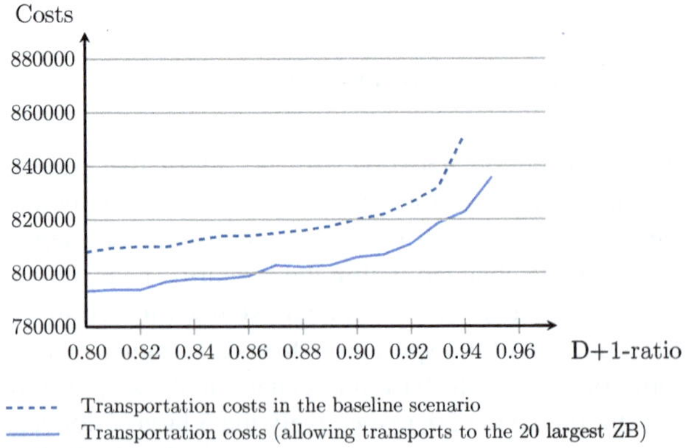

Fig. 8 Transportation costs with and without direct transportations

Table 2 Cost savings and service level improvements through direct transports in scenario 2

(a) Cost savings through direct transports	
D + 1-ratio in the baseline scenario (%)	Cost savings through direct transportations (%)
80	1.8
81	1.9
82	1.9
83	1.9
84	1.8
85	1.8
86	1.8
87	1.5
88	1.7
89	1.8
90	1.8
91	1.9
92	1.8
93	1.7
94	3.5
(b) Service level improvements	
D + 1-ratio in the baseline scenario (%)	Improvements of the D + 1-ratio in percentage points through direct transportations
90	3.54
91	2.87
92	2.26
93	1.68

in Fig. 8 are very similar to the results presented so far. But looking on it in more detail and also considering the computational results, it appears that an increase in shipments reinforces the positive effects of direct shipments. Table 2a exhibits the potential savings of direct shipments in case of an increased shipping volume in dependence on a given service level. In this scenario the average cost savings due to direct shipments are almost 2 %. According to that an increase in shipping volumes by 10 % leads to an augmentation of the cost differences by 0.5 %.

Besides increasing cost savings, service levels can be increased even more than in the baseline scenario without causing additional costs. Table 2b shows the calculated increases of the D + 1-ratio in comparison to the baseline scenario considering an increased shipment quantity. On average, the D + 1-ratio in the range of 90 % to 93 % can be increased by 2.59 percentage points, whereas the improvement in the same range in scenario 1 is only 2.17 percentage points.

These results clarify that direct transports are getting even more economical and profitable by an increase in shipping volumes. Increasing volumes lead to a stronger utilization of sorting capacities in hubs and, thus, existing bottlenecks will be reinforced by additional shipping quantities. Moreover, the increase in shipping volumes cause a higher utilization of transportation capacities on direct transportation links.

6 Conclusion

Customers in the German parcel market increasingly expect high service levels and, thus, short delivery lead times. Therefore, a project at Deutsche Post DHL was carried out to identify ways of enhancing the current service level within the German parcel network. Motivated by this project the paper analyzes the potential of alternative, direct transports to enhance the service level. In this context, a direct transport skips a hub in the network. To evaluate the potential of direct transports a mixed-integer linear program is formulated. The program minimizes the cost incurred by operating the transportation network while meeting a predetermined service level constraint.

The biggest challenge for the model formulation arises from the explicit consideration of the notion of time. Yet, without considering time within the network the $D+1$ service level constraint could not be adequately formulated. Though, to reduce the problem complexity direct transportation is limited to linking outbound hubs with delivery depots in the network. Furthermore, the solution space is reduced by limiting transports to practically feasible departure and or arrival times as well as origin-destinations pairs. After all, reducing the problem complexity allows for the model to be implemented in AIMMS and to be solved with CPLEX for real-life sized instances.

The results presented in Sect. 5 show that introducing direct transports to a parcel distribution network yield significant potential. For the underlying problem instance, transportation costs are reduced by 1.5 % on average for a fixed service level. When instead fixing the budget for transportation costs service levels are increased by 0.83 % on average. At the same time, the results support the assumption that the marginal benefit of additional direct transports is decreasing. A significant share of the identified potential is based on direct transports to the biggest delivery depots in Germany. When decreasing the required size of directly connected delivery depots the potential reduction of transportation costs decreases at the same time. Furthermore, the positive cost effect of direct transports increases with a rising number of shipped items. Assuming a 10 % increase of shipments, transportation costs can be lowered by 2 % in comparison to a scenario with no direct transports. Thus, direct transportations seem to be an appropriate measure to reduce transportation costs within a parcel distribution network while meeting predefined service levels. Based on the results of the mentioned project at Deutsche Post DHL, the company is currently testing direct transports from some outbound hubs to new build mechanized delivery depots.

Although the results of the optimization are quite satisfactory, the computational times of the problem are not. Further research will be done to develop a heuristical solution method and to compare the quality of the heuristical solution with the quality of the solution obtained with CPLEX. Providing CPLEX with a start solution could be another approach for reducing the computational times of the problem.

References

1. (2012) AIMMS: Paragon decision technology. http://www.aimms.com
2. Armacost AP, Barnhart C, Ware K, Wilson A (2004) UPS optimizes its air network. Interfaces 34(1):15–25
3. Barnhart C, Krishnan N, Kim D, Ware K (2002) Network design for express shipment delivery. Comput Optim Appl 21(3):239–262
4. Barnhart C, Schneur RR (1996) Air network design for express shipment service. Oper Res 44(6):852–863
5. Cheung W, Leung LC, Wong YM (2001) Strategic service network design for DHL Hong Kong. Interfaces 31(4):1–14
6. Christiansen M, Fagerholt K, Nygreen B, Ronen D (2007) Maritime transportation. Handb Oper Res Manag Sci 14:189–284
7. Christiansen M, Fagerholt K, Ronen D (2004) Ship routing and scheduling: status and perspectives. Transp Sci 38(1):1–18
8. Cordeau JF, Toth P, Vigo D (1998) A survey of optimization models for train routing and scheduling. Transp Sci 32(4):380–404
9. CPLEX: IBM ILOG (2012) http://www.ibm.com
10. Crainic TG (1999) Long-haul freight transportation. Handb Transp Sci 56:433–491
11. Crainic TG (2000) Service network design in freight transportation. Eur J Oper Res 122(2): 272–288
12. Crainic TG, Kim KH (2007) Intermodal transportation. Handb Oper Res Manag Sci 14: 467–537
13. Crainic TG, Laporte G (1997) Planning models for freight transportation. Eur J Oper Res 97(3):409–438
14. Grünert T, Sebastian HJ (2000) Planning models for Long-haul operations of postal and express shipment companies. Eur J Oper Res 122(2):289–309
15. Irnich S (2002) Netzwerk-Design für zweistufige Transportsysteme und ein Branch-and-Price-Verfahren für das gemischte Direkt- und Hubflugproblem. Ph.D. thesis, Lehr- und Forschungsgebiet Operations Research und Logistik Management, RWTH Aachen
16. Kim D (1997) Large scale transportation service network design: models, algorithms and applications. Ph.D. thesis, Department of Civil and Environmental Engineering at the Massachusetts Institute of Technology
17. Kim D, Barnhart C, Ware K, Reinhardt G (1999) Multimodal express package delivery: a service network design application. Transp Sci 33(4):391–407
18. Kuby MJ, Gray RG (1993) The hub network design problem with stopovers and feeders: the case of federal express. Transp Res A-Policy 27:1–12
19. MRU GmbH (2009) Primärerhebung auf den Märkten für Kurier-, Express- und Paketdienste
20. Statistisches Bundesamt Deutschland (2010) Verbraucherpreisindex für Deutschland. http://www-genesis.destatis.de/genesis/online
21. Vahrenkamp R (2007) Logistik—Management und Strategien, 6., überarb. und erw. Aufl. edn. Oldenbourg, München

New Approaches of Realizing an Optimized Network

Julia Hillebrandt

Abstract Hub location is a strategic management decision with far-reaching consequences for a company. Designing the hub network and the positions of the locations are critical components according the delivery of the logistic system. Many of the extant research deals with this issue. Yet, even if there is a network, it needs to be continuously adapted to the different conditions. Transportation quantity can increase the degree of capacity of the logistic system, which, consequently, ascends the current network load. The abilities of the network have to be reviewed and must be adjusted if necessary. As a result of this, there will be a need to do a re-optimization intermittently. This optimization is not finished after solving the problem. The research questions are as follows: how to transform the network in the optimized network? How to realize this improved adjustment? This paper deals with the topic of optimizing adoption process: how to realize an optimized network with optimization methods.

1 Introduction

As a result of the increase in trading via the Internet, the requirements of the parcel network continue to evolve. To be prepared for future conditions, there must be an ongoing optimization of the network. It continues to be a challenge to efficiently transport the parcels. The parcel network is a hub location network with sorting capacity at the different hubs.

If we assume, that the degree of packages will grow year over year in the next years, there will be a need for a new sorting capacity. The development process of the parcel network will spend much time. To minimize the risks to the network owner, the change process must be made more comfortable for the owner. It is a great advantage to realize such a target as early as possible to get experience with the network configuration. This, at least, will guarantee a quality baseline of the network and minimize disassembly costs. Hub location has been addressed by numerous papers

J. Hillebrandt (✉)
Deutsche Post DHL, Charles-de-Gaulle Straße 20, 53113 Bonn, Germany
e-mail: julia.hillebrandt@deutschepost.de

H.-J. Sebastian et al. (eds.), *Quantitative Approaches in Logistics and Supply Chain Management*, Lecture Notes in Logistics,
DOI 10.1007/978-3-319-12856-6_7

in the literature. O'Kelly [1] presents the first mixed integer linear programming formulation for the hub location problem. He showed that the problem is NP-hard, and he suggested two heuristics for solving it. Based on the issue of this work, a wide research area has resulted. An overview of hub location networks is provided by Campbell [2], O'Kelly and Miller [3], and Skorpion-Kapov et al. [4]. Hubs serve as transshipment points and allow indirect connections between sinks and sources. The result is a hub-and-spoke network, in which flow between any origin and destination can only take place through the hubs. In a hub network we distinguish between single and multiple assignments depending on connection between hubs and other locations. In the single assignment, each sink or source is connected to exactly one hub, in comparison to the multiple assignments, where every sink or source could connect to more than one hub. The aim of a hub location network is to minimize the complete transportation cost. There is a wide range of applications for the hub location problem, including airline passenger flows, communication traffic, and package delivery networks. Owing to these different areas of application, the models have to deal with a huge amount of data. Yaman [19] explains allocation strategies in hub networks. Thus, plenty of approaches have been developed for solving hub location problems, such as problem-specific heuristics [1], tabu search [5], hybrids of genetic algorithms and tabu search [6], and neural networks [7]. But the question is: What happens after this optimization? How can we adapt an already established network to new requirements? Some research papers have been published which handle the issue of expansion as a hub location problem [8, 9]. These papers describe the concept of reassigning capacities between hubs. Furthermore, Luss [10] also presents a survey of capacity expansion network. Campbell [18] describes hub location for time definite transportation. But this approach only deals with the description of which size a hub should be extended. But none of these papers answer the question of how to realize such a network after optimization. The question guiding this research is, when the best time for the expansion would be. What happens to the sinks and sources which are assigned to the hub during construction? Will they connect to other hubs? This paper deals with this specific topic. The aim of this paper is to present new approaches to realizing an optimized network with optimization models, which describe the adoption process. It focuses on a single assignment hub location network. Therefore, there are two different ideas—the big bang method and an iterative approach. The main ideas will be presented in the next section.

2 New Approaches to Realizing an Optimized Network

In this paper, a special hub location problem will be studied. An example of this hub network with three hubs is given in Fig. 1. Every item starts from a source to an addressed sink. The task is to minimize the transportation costs. The difference of a usual used hub location network is, that a sink can't be a source. Furthermore the items are addressed and can be clearly located. The baseline of this used network is the parcel network as described in Müller and Hillebrandt [17]. The transported

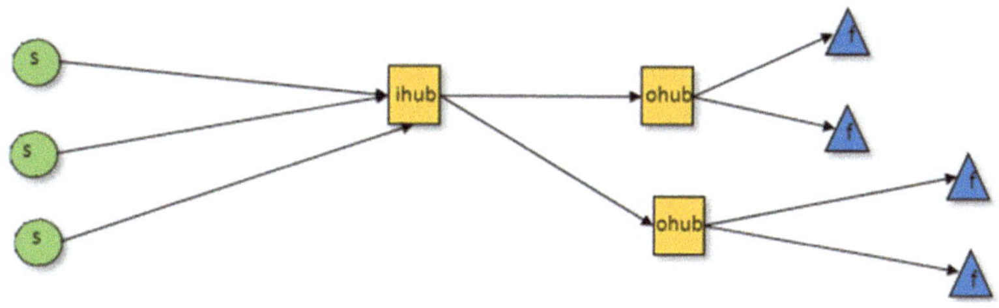

Fig. 1 An example of a hub network

items are parcels. The network is divided into three parts including a pre-, a main and a post-sector. The following section describes the network in more detail.

In the first sector, the quantity flows from the sources to an inbound hub (ihub). This hub works as a sorting center. It consolidates the quantity with the same outbound hub (ohub). After consolidation it will transferred to an outbound hub and then delivered to the different sinks. The outbound hub sorts the quantity to the different addressed sinks. This network was optimized in Müller and Hillebrandt [17], the result being, that there are hubs which should be closed and there are hubs which should be expanded. The question is, then, what is the best way to do this? What is the best time to rebuild a hub? The functionality and the safeguarding of the transport chain have to be permanently and lastingly guaranteed, so that there is a good service quality. Service quality means to send an item through the three parts (pre-, main, und post-sector) of the network within defined time. This is a service standard for the customers, who use this service for sending their items. Otherwise it means a reduction of service quality and a potential of losing clients. The presented model will be used to get a project schedule. It answers the question of where the different sinks and sources will be assigned during the construction.

To implement the optimization to obtain a project schedule, two approaches were identified. The main ideas are based on the concepts and strategies of software engineering. There are two different approaches to realizing a network optimization of existing software systems that are mentioned in literature, iterative approach and big bang development [11, 12]. The strategy of iterative and incremental development of software describes an iterative switching of individual components over time. The old and new systems work simultaneously for a period of time, easing the transition. One advantage of the iterative approach is the lower implementation risk. The user can be trained according the project's progress. This leads to new experiences being gained. However, this approach requires a longer time horizon until the software is fully operational and the data must be processed with each step in comparison to other methods. Using the big-bang method means to implement the software with all the components at a predetermined time. The existing system is replaced immediately. The final state is reached in a straightforward way in comparison to the other strategy. A disadvantage of the big bang approach is a high implementation risk, owing to the short realizing time horizon. The flexibility is very limited. A tight project management is required to ensure a smooth processing within the system.

Both concepts should be used for the realization of an optimized network. The basic ideas of the iterative approach are not directly transferable. The sorting capacity is steadily expanded. The fixed costs of inbound and outbound sorting centers are very high. Therefore, replacing a nearby sorting center is not common; it would be too expensive to realize a one-by-one iterative approach. Consequently, the extension is realized only in a sorting center. The sorting center should be extended until the structural limitations of buildings and materials. The hubs could sort more quantity and yet, more sinks and sources could be assigned, which would lead to cost reduction at the network and a better handling with increasing quantity. Only the sorting centers which help to improve the network load should be extending. The aim is to get a quality and cost-effective network. An advantage of this approach is that the capacity of each sorting center can be directly adapted to the specific parcel volume. If the amount has not been widely developed, the expansion can be deferred over a period of time. The sorting center could sort continuously and may not be closed for a longer period. This leads to minimal restrictions on the quality of service. An adequate extension cannot be guaranteed due to physical circumstances, such as an installation of additional sorting lines. Moreover, not all existing systems can be extended and will not always enable adjustments. In addition, a change in capacitance includes all hardware and software components change at the same time.

With the big bang method the sorting centers will be expanded to their final size after closing the sorting center. During the renewal of the sorting center a reduction in the quality of the network is accepted. The sinks and sources are assigned to another open sorting center during the extension. The open sorting centers take on the amount of the expanding sorting center in order to sort it. After the reopening, reallocations ensue. The costs decrease and the quality could increase for the complete network. First, those sorting centers should be closed and expanded with the most significant cost savings. An advantage of this approach is that an old sorting facility will be completely replaced during a short time horizon. There is only one working and operating system. The optimized network will be realized faster in comparison to other strategies. The operation of the facility during the renovation is completely shut down. No special arrangements need be taken during the reconstruction phase in the sorting center. A disadvantage of this method is the reduction in service quality during this construction phase. The process of reconstruction always entails under difficult conditions. It requires a strict project management.

2.1 Big Bang Method

To formulate the problem, the following notations are defined:

I	set of inbound sorting centers ($i \in I$)
J	set of outbound sorting centers ($j \in J$)
F	set of sinks ($f \in F$)
S	set of sources ($s \in S$)
I_o	set of closing inbound sorting centers ($i_o \in I_o$)

J_o set of closing outbound sorting centers ($j_o \in J_o$)
L status l of a sorting center: 1: renovation 2: reopened ($l \in L$)
T time periods($t \in T$)
$b^t_{s,f}$ supply of source s with destination f at the time t
be^t_f demand of sink f at the time t
$c^1_{s,i}$ cost per item between source s and inbound sorting center i
$c^2_{i,j}$ cost per item between inbound sorting center i and outbound sorting center j
$c^3_{j,f}$ cost per item between sink f and outbound sorting center j at
am^t_s amount of source s at time t
UA_i rebuild time of the inbound sorting center i
UE_j rebuild time of the outbound sorting center j
OA_t fixed cost for opening at time t an inbound sorting center that is to be closed
OE_t fixed cost for opening at time t an outbound sorting center that is to be closed
$K^t_{i,l}$ capacity of an inbound sorting center i at the level l at time t
$Kn^t_{i,l}$ capacity of an inbound sorting center j at the level l at time t
$KO^t_{j,l}$ capacity of an outbound sorting center j at the level l at time t
SF factor of increasing
KQ rate of capacity
CK cost savings of consolidation

Decision variables:

$$x^t_{s,i} \begin{cases} 1, & \text{if a source s is located to an inbound sorting center i} \\ & \text{at time t} \\ 0, & \text{otherwise} \end{cases}$$

$$p^t_{j,f} \begin{cases} 1, & \text{if a sink f is located to an outbound sorting center i} \\ & \text{at time t} \\ 0, & \text{otherwise} \end{cases}$$

$$y^t_{i,o} \begin{cases} 1, & \text{if a closing inbound sorting center i is open at time t} \\ 0, & \text{otherwise} \end{cases}$$

$$ye^t_{j,o} \begin{cases} 1, & \text{if a closing outbound sorting center i is open at time t} \\ 0, & \text{otherwise} \end{cases}$$

$$ka^t_{i,l} \begin{cases} 1, & \text{if the status l of the inbound sorting center i is reached at time t} \\ 0, & \text{otherwise} \end{cases}$$

$$ke^t_{j,l} \begin{cases} 1, & \text{if the status l of the outbound sorting center j is reached at time t} \\ & \text{time t} \\ 0, & \text{otherwise} \end{cases}$$

$k^{f,t}_{i,j}$ flow between inbound sorting center i and outbound sorting center j with destination f at time t
$KA_{t,i}$ capacity adjustment factor for inbound sorting center i at time t
$KP_{t,j}$ capacity adjustment factor for outbound sorting center j at time t

The problem can be formally stated as follows:

$$\text{Min} \quad \Sigma_s \Sigma_i \Sigma_t c^1_{s,i} * x^t_{s,i} * am^t_s + CK * \Sigma_i \Sigma_j \Sigma_t \Sigma_f c^2_{i,j} * k^{f,t}_{i,j} + \Sigma_j \Sigma_f \Sigma_t c^3_{i,j} * p^t_{j,f} * be^f_t + \Sigma_{i_o} \Sigma_t OA_t * y^t_{i_o} + \Sigma_{j_o} \Sigma_t OE_t * ye^t_{j_o} \tag{1}$$

$$\Sigma_s b^t_{s,i} * x^t_{s,i} = \Sigma_j k^{f,t}_{i,j} \qquad \forall t \in T, f \in F, i \in I \tag{2}$$

$$\Sigma_i k^{f,t}_{i,j} = be^t_f * p^{f,t}_{i,j} \qquad \forall t \in T, f \in F, j \in J \tag{3}$$

$$\Sigma_j p^t_{j,f} = 1 \qquad \forall t \in T, f \in F \tag{4}$$

$$\Sigma_s x^t_{s,i} = 1 \qquad \forall t \in S, t \in T \tag{5}$$

$$\Sigma_i K_{i,1} - K_{i,1} * ka^t_{i,1} + ka^t_{i,2} * K_{1,2} - \Sigma_{i_o} Ki_{o,1} * y^t_{i_o} \geq KQ * \Sigma_f b^t_f + \Sigma_i KA_{t-1,i} \qquad \forall t \in T, t > 1 \tag{6}$$

$$\Sigma_j Kn_{j,1} - Kn_{j,1} * ke^t_{j,1} + ke^t_{j,2} * Kn_{j,2} - \Sigma_{j_o} Kn_{j_o,1} * ye^t_{j_o} \geq KQ * \Sigma_f b^t_f + \Sigma_j KP_{t-1,j} \qquad \forall t \in T, t > 1 \tag{7}$$

$$\Sigma_s x^t_{s,i} \leq y^t_{i_o} * M \qquad \forall i_o \in I_o, t \in T, i \in I, i = i_o \tag{8}$$

$$\Sigma_f p^t_{f,j} \leq ye^t_{j_o} * M \qquad \forall j_o \in J_o, t \in T, j \in J, j = j_o \tag{9}$$

$$y^t_{i_o} \geq y^{t+1}_{i_o} \qquad \forall i_o \in I_o, t \in T \tag{10}$$

$$y^t_{io} = ye^t_{j_o} \qquad \forall i_o \in I_o, j_o \in j_o, i_o = j_o, t \in T \tag{11}$$

$$\Sigma_s \Sigma_f b^t_{s,f} * x^t_{s,i} \leq K_{i,1} - K_{i,1} * ka^t_{i,1} + ka^t_{1,2} * K_{i,2} + KA_{t,i} \qquad \forall t \in T, i \in I \tag{12}$$

$$\Sigma_f be^t_f * p^t_{f,j} \leq Kn_{j,1} - Kn_{j,1} * ke^t_{j,1} + ke^t_{j,2} * kn_{j,2} + KP_{t,j} \qquad \forall t \in T, j \in J \tag{13}$$

$$ka^t_{i,2} = ka^{t-UA_i}_{i,1} \qquad \forall t \in T, i \in I \tag{14}$$

$$ke^t_{j,2} = ke^{t-UA_i}_{j,1} \qquad \forall t \in T, j \in J \tag{15}$$

$$\Sigma_i KA_{t,i} \leq M * \Sigma_i ka^t_{i,1} - ka^t_{i,2} \qquad \forall t \in T \tag{16}$$

$$\Sigma_j KP_{t,j} \leq M * \Sigma_j ke^t_{j,1} - ke^t_{j,2} \qquad \forall t \in T \tag{17}$$

$$KA_{t,i} \leq SF * (K_{i,1} - K_{i,1} * ka^t_{i,1} + K_{i,2} * Ka^t_{i,2}) \quad \forall t \in T, i \in I \tag{18}$$

$$KP_{t,j} \leq SF * (Kn_{j,1} - Kn_{j,1} * ke^t_{j,1} + K_{j,2} * ke^t_{j,2}) \quad \forall t \in T, j \in J \tag{19}$$

$$ka^t_{i,1} \geq ka^t_{i,2} \qquad \forall t \in T, i \in I \tag{20}$$

$$ke^t_{j,1} \geq ke^t_{j,2} \qquad \forall t \in T, j \in J \tag{21}$$

$$y^t_{i_o} \in \{0,1\}, ye^t_{j0} \in \{0,1\}, k^{f,t}_{i,j} \geq 0$$

$$ka^t_{i,l} \in \{0,1\}, ke^t_{j,l'} \in \{0,1\}, x^t_{s,i} \in \{0,1\}, p^t_{j,f} \in \{0,1\}$$

The objective function (1) minimizes the transportation costs, consisting of a pre-, a main and a post-sector and also the fix cost for opening a closing sorting center.

Constraint (2) ensures the flow conservation at the inbound sorting center. The whole amount of the pre-sector has to be the same like the amount of the main sector. This also applies accordingly for the constraint (3) for the outbound sorting center. Constraints (4) and (5) are single allocation constraints. This ensures that every source is allocated to an inbound sorting center and every sink to an outbound sorting center. The constraint (6) ensures that a specified percentage of capacity must be reserved for an unsorted quantity from previous periods. Only a limited loss of service quality is accepted. The restriction (7) describes this analogy for an outbound sorting center. If the sorting center is to be expanded, it should be closed first. Due to the closure of a sorting center, a certain amount cannot be sorted. This case occurs when the adjacent sorting centers also operate at limit and consequently there is no free additional capacity. This can lead to deterioration in service quality. The amount will be sorted in the subsequent period. This creates a trade-off between fast conversions of all sorting centers to get a lower-cost network rapidly and to avoid deterioration of the delivery quality during the expansion. Therefore, a percentage range has to be defined at the beginning of the model run, regarding the extent of the sorting capacity that is acceptable to lose during the expansion. So that during the expansion phase, no model restrictions are violated, an appropriate auxiliary variable has to be declared. Then the amount could collect in this variable for sorting during the next period.

A sink or source may not be assigned to a closed inbound or outbound sorting center (restrictions 8 and 9). When an inbound sorting center is closed, it is also closed in subsequent periods. This is ensured by the constraint (10). If an inbound sorting center is closed the same outbound center is also closed (constraint 11). A sorting center will only be assigned so much quantity as sorting capacity is present. Consequently, it has to be checked as to which condition the sorting center is in. There are three possibilities: the sorting center has its initial capacity, it is currently under construction, or it is has expanded to its final size. If the sorting center is under construction, no source or sink will be allocated to this sorting center. The situation is illustrated in the constraint (12). The same applies with the constraint (13) for the outbound sorting center. The constraint (14) ensures that during the extension of the inbound sorting center, a conversion period is observed. The same is illustrated by restriction (15) for the outbound sorting center. A capacity equalization variable may only use during the renovation of an outbound or inbound sorting center (restrictions 16 and 17). The use of capacity balancing will automatically lead to a worse delivery quality. In addition, the amount of a sorting center under construction is located on the adjacent sorting centers. Thus, only a certain, predetermined percentage for each sorting center is used for the compensation of sorting adjacent quantity. This is to prevent a single sorting center getting the entire shipment quantity of an upgraded sorting center. This is ensured by constraint (18). The same fact is guaranteed by constraint (19) for the outbound sorting center. If a sorting center has to be extended it, cannot be extended again (constraints 20 and 21).

2.2 Iterative Approach

To formulate the problem, the following notations are defined:

I set of inbound sorting centers (i ∈ I)
J set of outbound sorting centers (j ∈ J)
F set of sinks (f ∈ F)
S set of sources (s ∈ S)
I_o set of closing inbound sorting centers ($i_o \in I_o$)
J_o set of closing outbound sorting centers ($j_o \in J_o$)
T time periods(t ∈ T)
$b^t_{s,f}$ supply of source s with destination f at the time t
be^t_f demand of sink f at the time t
$c^1_{s,i}$ cost per item between source s and inbound sorting center i
$c^2_{i,j}$ cost per item between inbound sorting center i and outbound sorting center j
$c^3_{j,f}$ cost per item between sink f and outbound sorting center j
Y_t fixed cost for opening a to closing inbound sorting center at time t
K^t_i capacity of an inbound sorting center i at time t
Kn^t_j capacity of an outbound sorting center j at time t
$KS_{i,o}$ rate of capacity at the inbound sorting center i with level o
KP^o_j rate of capacity at the outbound sorting center j with level o
SF factor of increasing
CK cost savings of consolidation

$[p^t_{j,f}]$ Decision variables:

$x^t_{s,t}$ $\begin{cases} 1, \text{ if a source s is located to an inbound sorting center i at time t} \\ 0, \text{ otherwise} \end{cases}$

$p^t_{j,f}$ $\begin{cases} 1, \text{ if a sink f is located to an outbound sorting center j at time t} \\ 0, \text{ otherwise} \end{cases}$

$y^t_{i_o}$ $\begin{cases} 1, \text{ if a closing inbound sorting center i is open at time t} \\ 0, \text{ otherwise} \end{cases}$

$ye^t_{j_o}$ $\begin{cases} 1, \text{ if a closing inbound sorting center i is open at time t} \\ 0, \text{ otherwise} \end{cases}$

$kpa^t_{i,o}$ expansions level o of the inbound sorting I center at time t
$kpe^t_{j,o}$ expansions level o of the outbound sorting j center at time t
$KA_{t,i}$ capacity adjustment factor for inbound sorting center i at time t
$KP_{t,j}$ capacity adjustment factor for outbound sorting center j at time t
$k^{f,t}_{i,j}$ flow between inbound sorting center i and outbound sorting center j with destination f at time t

$$\text{Min} \quad \Sigma_s \Sigma_i \Sigma_t c^1_{s,i} * x^t_{s,i} + CK * \Sigma_i \Sigma_j \Sigma_t \Sigma_f c^2_{i,j} * k^{f,t}_{i,j} + \Sigma_j \Sigma_f \Sigma_t c^3_{i,j} * p^t_{j,f} * be^t_f + \Sigma_{i_o} \Sigma_t Y_t * y^t_{i_o} + \Sigma_{j_o} \Sigma_t * ye^t_{j_o} \tag{22}$$

$$\Sigma_s b^t_{s,f} * x^t_{s,i} = \Sigma_j k^{f,t}_{i,j} \qquad \forall t \in T, f \in F, i \in I \tag{23}$$

$$\Sigma_i k^{f,t}_{i,j} = be^t_f * p^{f,t}_{i,j} \qquad \forall t \in T, f \in F, j \in J \tag{24}$$

$$\Sigma_j p^t_{j,f} = 1 \qquad \forall t \in T, f \in F \tag{25}$$

$$\Sigma_i x^t_{s,i} = 1 \qquad \forall s \in S, t \in T \tag{26}$$

$$y^t_{i_o} \geq y^{t+1}_{i_o} \qquad \forall i_o \in I_o, t \in T \tag{27}$$

$$y^t_{i_o} = ye^t_{j_o} \qquad \forall i_o \in I_o, j_o \in J_o, i_o = j_o, t \in T \tag{28}$$

$$\Sigma_s x^t_{s,i} \leq y^t_{i_o} * M \qquad \forall i_o \in I_o, t \in T, i \in I \tag{29}$$

$$\Sigma_f p^t_{f,j} \leq ye^t_{j_o} * M \qquad \forall j_o \in J_o, t \in T, j \in J \tag{30}$$

$$\Sigma_s x^t_{s,i} \leq K^t_i + \Sigma_o KS_{i,o} * kpa^t_{i,o} + KA_{t,i} \quad \forall i_o \in I, t \in T \tag{31}$$

$$\Sigma_f p^t_{j,f} \leq Kn^t_j + \Sigma_o KP_{j,o} * kpe^t_{j,o} + KP_{t,j} \quad \forall j_o \in J, t \in T \tag{32}$$

$$\Sigma_i K^t_i + \Sigma_o KS_{i,o} * kpa^t_{i,o} - \Sigma_{i_o} K_{i_o,1} * y^t_{i_o} \geq KQ * \Sigma_f b^t_f + \Sigma_i kA_{t-1,i} \quad \forall t \in T, t > 1 \tag{33}$$

$$\Sigma_j Kn^t_j + \Sigma_o KP_{j,o} * kpe^t_{j,o} - \Sigma_{j_o} Kn_{j_o,1} * ye^t_{j_o} \geq KQ * \Sigma_f b^t_f + \Sigma_i kP_{t-1,j} \quad \forall t \in T, t > 1 \tag{34}$$

$$KA_{t,i} \leq SF * (\Sigma_i K^t_i + \Sigma_o KS_{i,o} * kpa^t_{i,o}) \quad \forall t \in T, i \in I \tag{35}$$

$$KP_{t,j} \leq SF * (\Sigma_j Kn^t_j + \Sigma_o KP_{j,o} * kpe_{j,o}t) \quad \forall t \in T, j \in J \tag{36}$$

$$y^t_{i_o} \in \{0,1\}, ye^t_{j_o} \in \{0,1\}, k^{f,t}_{i,j} \geq 0, x^t_{s,i} \in \{0,1\}, p^t_{j,f} \in \{0,1\}$$

The objective function (22) minimizes the transportation costs, consisting of a pre-, a main and post-sector. Constraint (23) ensures the flow conservation at the inbound sorting center. The whole amount of the pre-sector has to be the same as the amount of the main sector. This also applies accordingly for constraint (24) for the outbound sorting center. Constraints (25) and (26) are single allocation constraints. This ensures that every source is allocated to exactly one inbound sorting center and every sink to an outbound sorting center. When a sorting center is closed, it is also closed in subsequent periods (restrictions 27 and 28). Constraints (29) and (30) ensure that a sink or source cannot be allocated to a closed sorting center. Each sorting center has a capacity and can therefore only sort a limited quantity (constraints 31 and 32). However, the capacity can increase over time with a capacity increasing factor. Constraints (33) and (34) consider that a pre-defined service quality will be reached. It also ensures that the quantity of the previous time horizons also has to be sorted. In addition, the amount of a sorting center under construction is located on the adjacent sorting centers. Thus, only a certain, predetermined percentage for each sorting center is used for the compensation of sorting adjacent quantity. This is to prevent a single sorting center getting the entire shipment quantity of an upgraded sorting center. This is ensured by constraint (35). The same fact is guaranteed by constraint (36) for the outbound sorting center.

3 Computational Results

3.1 Test Data

A common data set for hub location problems, which is usually used for building and executing high effective tests concerning algorithms, has been already discussed by Fotheringham [13] and also by O'Kelly [1]. This paper cannot use this common data set, because it does not apply to capacity restrictions on the one hand; on the other hand, this paper will handle source and sink as two different locations. Furthermore, the respective paper does not include fixed costs. For testing the above-mentioned models some data has been generated. Therefore, *Gauss-Krüger* coordinates of towns and communes from all over Germany [14] have been collected. From this pool of data random coordinates have been selected for sources and sinks. It was taken, that the coordinates of sinks and sources are evenly distributed. The main idea was to review and check the introduced models. Additionally, 80 random potential sorting centers have been selected of this coordinates. From of the pool of 80 sorting centers, 8 were chosen and solved with a p-center model, which will be pictured later on. The distances between source and sorting center, and sink to sorting centers, have been calculated, using the Euclidean distance algorithm. Between sink and source, packages will be transferred and sorted in the sorting centers. The amount of packages is built as follows: To figure out which region in Germany has the highest gross domestic product, a statistic provided by the *Statistisches Bundesamt* [15] has been analyzed. The regions with the highest gross domestic products will be given the most spending power and hence the biggest amount of items for transportation. The data has been extracted using allocation methods which are in line with that gross domestic product.

3.2 Initial Solution

With this generated data of the 80 potential sorting center locations, 8 locations shall be evaluated, using a version of the p-center model. This result gives information about the optimal solution of the relation between source and sink to p-sorting centers to keep costs low. This result will be used as an initial solution for all models, which are mentioned in this paper. The way, the p-center model is used in this paper, will be described as follows.

To formulate the problem, the following notations are defined:

I	set of inbound sorting centers (i ∈ I)
J	set of outbound sorting centers (j ∈ J)
F	set of sinks (f ∈ F)
S	set of sources (s ∈ S)
m_{sf}	supply of source s with destination f
be_f	demand of sink f
$c^1_{s,i}$	cost per item between source s and inbound sorting center i
$c^2_{i,j}$	cost per item between inbound sorting center i and outbound sorting center j
$c^3_{j,f}$	cost per item between sink f and outbound sorting center j
p	quantity of open sorting centers

Decision variables:

$$x_{s,i} \begin{cases} 1, & \text{if a source s is located to an inbound sorting center i} \\ 0, & \text{otherwise} \end{cases}$$

$$l_{j,f} \begin{cases} 1, & \text{if a sink f is located to an outbound sorting center j} \\ 0, & \text{otherwise} \end{cases}$$

$$h_i \begin{cases} 1, & \text{if an inbound sorting center i is open} \\ 0, & \text{otherwise} \end{cases}$$

$$b_j \begin{cases} 1, & \text{if an outbound sorting center i is open} \\ 0, & \text{otherwise} \end{cases}$$

$l^f_{i,j}$ flow between inbound sorting center i and outbound sorting center j with destination f

$$\text{Min } \Sigma_s \Sigma_i c^1_{s,i} * m_{s,f} * x_{s,i} + \Sigma_i \Sigma_j \Sigma_f c^2_{s,i} * k^2_{i,j} + \Sigma_j \Sigma_f c^3_{s,i} * be_f * l_{j,f} \quad (37)$$

$$\Sigma_s m_{s,f} * x_{s,i} = \Sigma_j k_{i,j}^f \qquad \forall f \in F, i \in I \tag{38}$$

$$\Sigma_i k_{i,j}^f = be_f * l_{j,f} \qquad \forall f \in F, j \in J \tag{39}$$

$$\Sigma_i h_i = p \tag{40}$$

$$\Sigma_j l_{j,f} = 1 \qquad \forall f \in F \tag{41}$$

$$\Sigma_I x_{s,i} = 1 \qquad \forall s \in S \tag{42}$$

$$\Sigma_f l_{j,f} = M * b_j \qquad \forall j \in J \tag{43}$$

$$\Sigma_s x_{s,i} = M * h_j \qquad \forall i \in I \tag{44}$$

$$h_i = b_j \qquad \forall j \in J, i \in I, i = J \tag{45}$$

$$x_{s,i} \in \{0, 1\}, l_{j,f} \in \{0, 1\}, h_i \in \{0, 1\}, b_j \in \{0, 1\}, k_{i,j}^f \geq 0$$

The objective function (37) minimizes the transportation costs, consisting of a pre-, a main and post-sector. Constraint (38) defines the flow conservation at the inbound sorting center. The whole amount of the pre-sector must be equal to the amount of the main sector. It also applies to constraint (39) depending on the outbound sorting center. Constraint (40) determines the quantity of open sorting centers. Constraints (41) and (42) describe that every sink and source is allocated to just one sorting center. Constraints (43) and (44) explain that sink and source can only be allocated to an open sorting center. Constraint (45) shows, that the outbound sorting center will be open, if the inbound sorting center is open, too. Executing this model, we will receive the following result:

Figure 2 gives information about all used coordinates and the solution of the model with the respective 8 chosen sorting centers. The light and dark green dots picture the sink and the sources. The light blue crosses mark the potential sorting centers. The red quadrates present the solution of the model.

3.3 Computational Results

With the models mentioned in this paper, the network expansion will be supported. This support can be made in two different ways, using the big bang methods and using the iterative approach. Therefore, using this generated data, the two models have been used to obtain results, which are compared with each other. Furthermore,

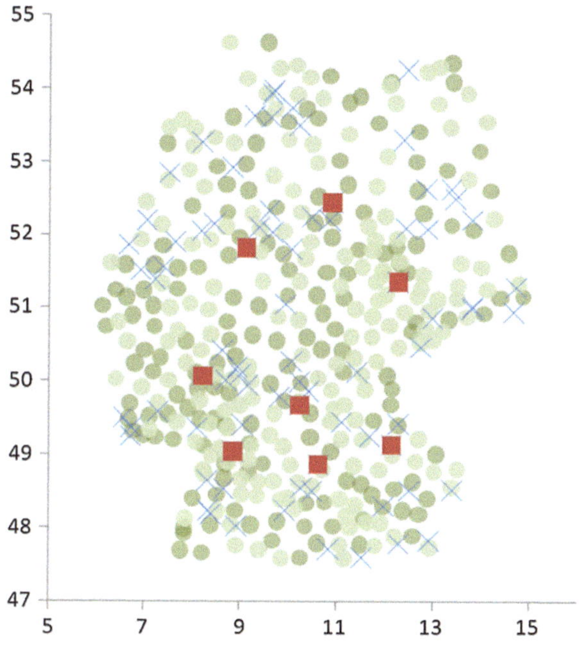

Fig. 2 Initial solution with sink and sources and potential sorting centers

several parameter settings have been made. Different cases involving the number of expanded sorting centers, the learning rate of costs between the sorting centers and the service quality in the form of available sorting capacity have been dealt with. The cases are calculated at an Intel Xeon double core 3.5 GHz processor with 48 GB RAM. The models are implemented in AIMMS [15] and solved with CPLEX [16] to prove and evaluate the results. Every case has a computing time up to 200,000 sec. The first calculation was using the big bang method. For the big bang method the cost of rebuilding is needed. In this example every rebuild takes 2 time units. The results are attached in the appendix. The following section discusses one result using the big bang method in more detail. The main question of all these models is to allocate the sources and the sinks to the sorting center to the sorting center during the reconstruction.

We will now describe one of the attached cases more in detail. In the example, one sorting center should be closed permanently and three sorting centers should be rebuilt. The service level should be 80 % of sorting capacity and 90 % learning rate between the sorting centers. Learning rate means that the higher amount in the inbound sorting centers from the sources leads to cost saving effects for the transportation between the sorting centers. The time horizon is about 4 time units. These are the defined parameters. The sorting centers were assigned numbers for identification. Figure 3 shows the allocation in the pre-sector during the different time horizons. The model result is to rebuild the sorting centers 12,565, 10,256,

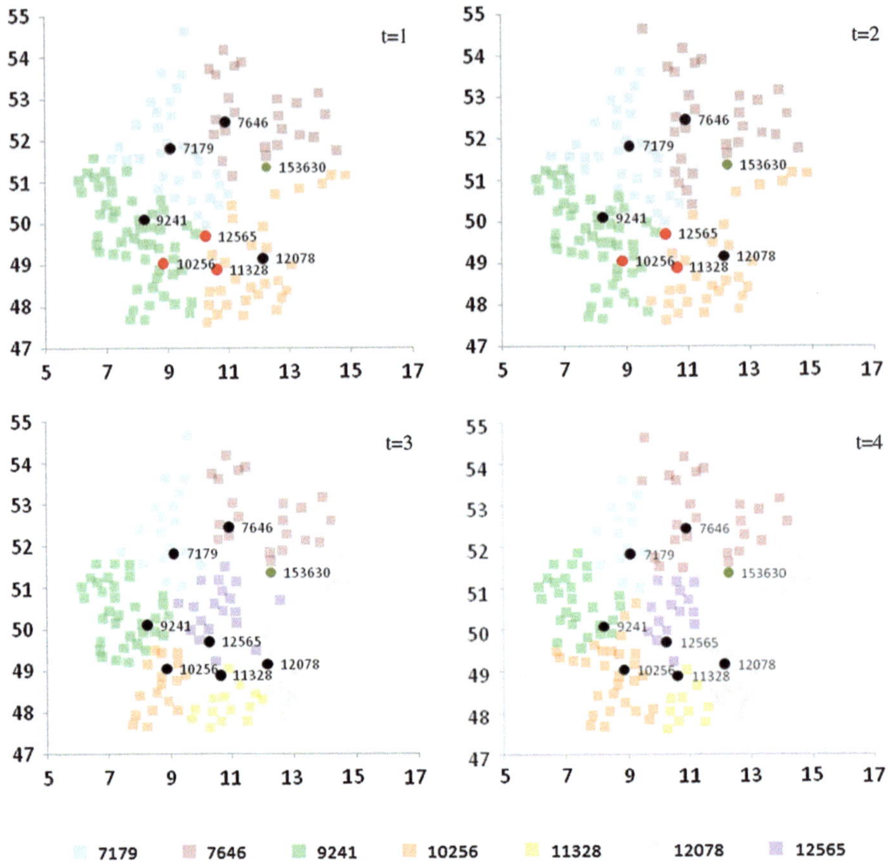

Fig. 3 Allocation of the sources to the sorting center during the four time periods with the big bang method

and 11,328. There are rebuild during the first and the second time unit (red points). Sorting center 153,630 should be closed permanently (dark green point). The sorting centers 7,646, 7,179, 9,241, and 12,078 take the amount of the closed sorting center during the expansion. In the post-sector the rebuild sorting centers are the same, of course. The main function costs are about 58,114,397. If you compare the different results, the overall cost are lower, if you only rebuild 2 sorting centers in comparison to rebuild 4 sorting centers. Therefore, the model was calculated with the specified, different parameters. The relevant results are attached.

Furthermore, the results of the iterative approach are also attached. Figure 4 shows one result more in detail. It describes the allocation of the sources in the pre-sector. The time horizons for the increase of sorting centers are 4 time units. The red points are the increased sorting centers and the dark green point is the permanently closed sorting center. The parameters are the same as in the big bang method. The overall costs are about 46,358,753. In this example, the savings of using the iterative approach

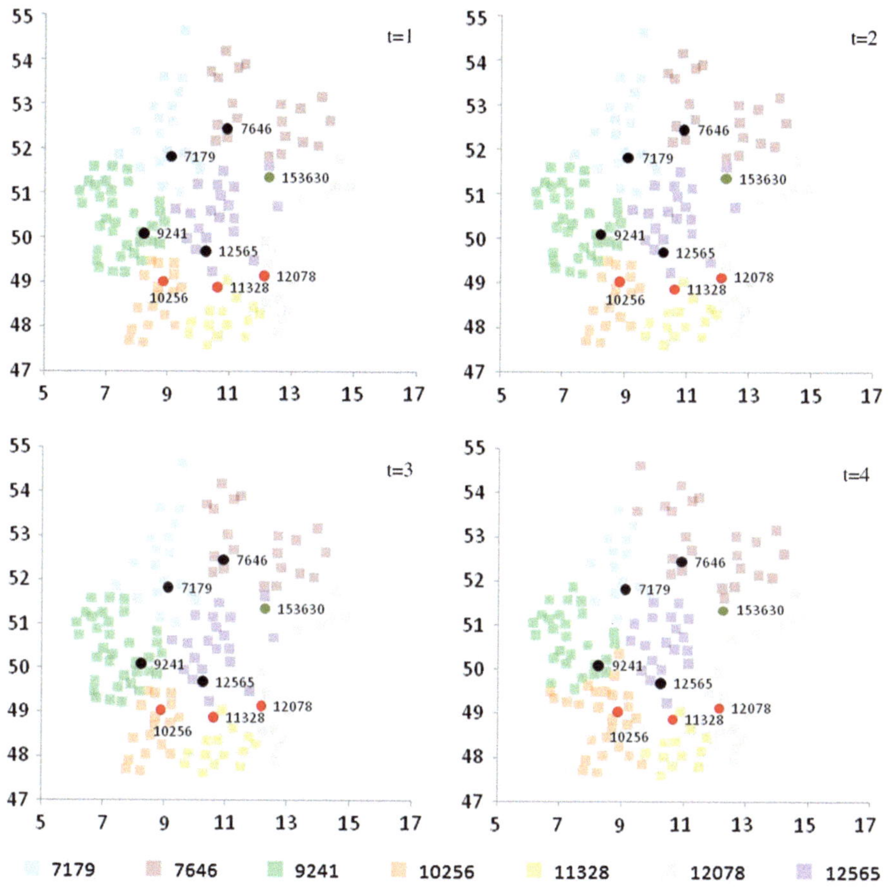

Fig. 4 Allocation of the sources to the sorting center during the four time periods with the iterative approach

is about 20 %. The major distinction of the iterative approach is that the sources and sinks could be allocated during the expansions process. There are no indirect routes.

3.4 Comparisons of Models

The basic structure of the iterative and the big bang approach are the same. There are allocation variables and overall transportation costs. Furthermore, there are hubs to close and reopen. The question is, when this rebuild should be. The main idea is to send an addressed parcel from a source to a sink. The difference is the expansion of the sorting capacity. On the one hand, with the iterative approach the capacity increases step-by-step. On the other hand, the big bang method makes the sorting capacity quickly available. If the big bang method and the iterative approach will be compared with each other, the overall costs of the iterative approach are much cheaper. With the model parameters there is a saving about 15–21% of the cost in comparison to the big bang method. The savings depend on the chosen parameters. It was not possible to get a result of all parameter settings with the big bang method, because some parameters are not solvable. Furthermore, both models rebuild nearly the same sorting centers. The most cost savings of the iterative approach is that there are no changes of routes for the parcel transportation. As long as there is enough sorting capacity for the amount, the iterative approach should be the cheapest way for a network extension. The question is how to realize such an iterative approach in practice. There are no infinite expansions possible, because of physical restriction of buildings and machines. The big bang methods in combination with the iterative approach are more in common in a practicable use. This should be evaluated in further papers.

4 Conclusion

This paper considers two approaches to realize an optimized network. The results allow a comparison of optimal parcel network solutions. The results of the models illustrate the different allocation of sources and sinks to the sorting centers during the time horizons. The comparison of the models shows that the overall costs of the iterative approach are much lower with the chosen parameters. The savings are about 15–21 only one or two hubs in comparison to rebuilding more hubs during a defined time. The reason is that the transportation costs are much higher during the rebuilding process, than the advantage of the rebuild sorting centers. Furthermore, the model was tested for small instances; if problem sizes become large, efficient heuristic procedures are necessities for solving these large problems. The two approaches need to be tested for their practical feasibility as well.

Attachment

See Tables 1 and 2.

Table 1 Results of using the big bang method

Case	Rate of learning (%)	#Rebuild	Rebuild	Service quality (%)	Objective function	Gap	#Variable	#Iteration
Case_1	1	2	9,241; 11,328	0.7	58,293,440.33	4.71	11,260	4,201,351
Case_2	1	2	9,241; 12,565	0.8	58,093,729.99	4.35	11,260	2,711,222
Case_3	1	2	9,241; 12,565	0.9	58,194,635.22	4.52	11,260	2,762,634
Case_4	1	2	10,256; 11,328	1	54,796,394.85	4.04	11,260	4,676,124
Case_5	1	3	7,646; 10,256; 11,328	0.7	58,065,360.4	4.27	11,260	3907311
Case_6	1	3	9,241; 11,328; 153,630	0.8	58,375,569.08	4.81	11,260	3,770,824
Case_7	1	3	12,078; 12,565; 153,630	0.9	58,717,308.16	5.23	11,260	3,620,855
Case_8	1	3	0	1	0	0.00	11,260	0
Case_9	1	4	10,256; 11,328; 12,565; 153,630	0.7	58,351,451.39	0.25	11,260	14,384,165
Case_10	1	4	10,256; 11,328; 12,565; 153,630	0.8	58,423,541.22	0.36	11,260	31328063
Case_11	1	4	0	0.9	0	0.00	11,260	7037
Case_12	1	4	0	1	0	0.00	11,260	0
Case_13	0.9	2	9,241; 12,565	0.7	58,080,502.01	3.89	11,260	5,944,314
Case_14	0.9	2	9,241; 11,328	0.8	57,874,392.88	3.46	11,260	5,836,869
Case_15	0.9	2	9,241; 12,565	0.9	58,218,102.21	4.16	11,260	5,858,592
Case_16	0.9	2	12,078; 12,565	1	58,619,785.44	3.43	11,260	5,657,346
Case_17	0.9	3	9,241; 11,328; 153,630	0.7	57,855,763.56	3.42	11,260	6119586
Case_18	0.9	3	10,256; 11,328;12,565	0.8	58,431,521.52	4.52	11,260	7,637,935

(continued)

Table 1 (continued)

Case	Rate of learning (%)	#Rebuild	Rebuild	Service quality (%)	Objective function	Gap	#Variable	#Iteration
Case_19	0.9	3	7179; 10,256; 153,630	0.9	58,185,070.51	3.56	11,260	7,009,901
Case_20	0.9	3	0	1	0	0.00	11,260	0
Case_21	0.9	4	10,256; 11,328; 12,565; 153,630	0.7	58,357,854.4	0.88	11,260	23,407,179
Case_22	0.9	4	10,256; 11,328; 12,565; 153,630	0.8	58,429,638.43	0.12	11,260	38,939,752
Case_23	0.9	4	0	0.9	0	0.00	11,260	7,724
Case_24	0.9	4	0	1	0	0.00	11,260	0
Case_25	0.8	2	9,241; 12,565	0.7	57,984,649.22	3.47	11,260	7939124
Case_26	0.8	2	9,241; 11,328	0.8	57,866,031.74	3.09	11,260	7,588,182
Case_27	0.8	2	7,179; 10,256	0.9	57,846,644.38	3.19	11,260	8,182,571
Case_28	0.8	2	10,256; 11,328	1	54,786,480.18	3.20	11,260	8,540,166
Case_29	0.8	3	7,646; 10,256; 11,328	0.7	57,821,963.88	2.99	11,260	9,118,980
Case_30	0.8	3	9,241; 11,328; 153,630	0.8	57,930,714.73	3.17	11,260	8,974,454
Case_31	0.8	3	7,179; 10,256; 153,630	0.9	58,114,396.62	2.67	11,260	9,160,546
Case_32	0.8	3	0	1	0	0.00	11,260	0
Case_33	0.8	4	10,256; 11,328; 12,565; 153,630	0.7	58,364,432.29	0.06	11,260	27,231,947
Case_34	0.8	4	10,256; 11,328; 12,565; 153,630	0.8	58,439,733.74	0.07	11,260	47,115,358
Case_35	0.8	4		0.9	0	0.00	11,260	8343
Case_36	0.8	4		1	0	0.00	11,260	0

Table 2 Results of using the iterative approach

Case	Rate of learning	#Rebuild	Rebuild	Service quality	Objective function	Gap	#Variable	#Iteration
Case_1	1	2	12,565; 9,241	0.7	46,543,967.59	0.28	11,316	7,208,768
Case_2	1	2	12,565; 9,241	0.8	46,546,336.06	0.29	11,316	11,168,527
Case_3	1	2	12,565; 9,241	0.9	46,548,154.55	0.29	11,316	6,895,543
Case_4	1	2	12,565; 9,241	1	46,637,817.82	0.43	11,316	7,782,298
Case_5	1	3	12,565; 10256; 11,328	0.7	46,286,641.42	0.10	11,316	11,265,546
Case_6	1	3	12,565; 10256; 11,328	0.8	46,283,280.28	0.09	11,316	11,329,193
Case_7	1	3	12,565; 10256; 11,328	0.9	46,283,280.28	0.09	11,316	12,250,260
Case_8	1	3	12,565; 10,256; 11,328	1	46,283,280.28	0.08	11,316	12,544,868
Case_9	1	4	12,565; 10,256; 11,328; 7,179	0.7	46,231,679.53	0.02	11,316	21,713,307
Case_10	1	4	12,565; 10,256; 11,328; 7,179	0.8	46,231,679.53	0.01	11,316	22,488,824
Case_11	1	4	12,565; 10,256; 11,328; 7,179	0.9	46,231,679.53	0.10	11,316	28,163,978
Case_12	1	4	12,565; 10,256; 11,328; 7,179	1	46,231,679.53	0.00	11,316	20,477,075
Case_13	0,9	2	12,565; 10,256	0.7	46,580,071.98	0.19	11,316	14,426,473
Case_14	0,9	2	12,565; 10,256	0.8	46,579,509.63	0.15	11,316	19,262,904
Case_15	0,9	2	12,565; 10,256	0.9	46,579,509.63	0.15	11,316	11,977,783
Case_16	0,9	2	12,565; 10,256	1	46,661,467.88	0.25	11,316	10,132,892
Case_17	0,9	3	12,565; 10,256; 11,328	0.7	46,319,374.44	0.00	11,316	10,829,257
Case_18	0,9	3	12,565; 10,256; 11,328	0.8	46,319,374.44	0.00	11,316	10,823,568
Case_19	0,9	3	12,565; 10,256; 11,328	0.9	46,319,374.44	0.00	11,316	11,354,380
Case_20	0,9	3	12,565; 10,256; 11,328	1	46,319,374.44	0.00	11,316	24,575,559
Case_21	0,9	4	12,565; 10,256; 11,328; 7,179	0.7	46,265,816.6	0.00	11,316	1,884,432
Case_22	0,9	4	12,565; 10,256; 11,328; 7,179	0.8	46,265,816.6	0.00	11,316	2,454,792

(continued)

Table 2 (continued)

Case	Rate of learning	#Rebuild	Rebuild	Service quality	Objective function	Gap	#variable	#Iteration
Case_23	0,9	4	12,565; 10,256; 11,328; 7,179	0.9	46,265,816.6	0.00	11,316	5,874,883
Case_24	0,9	4	12,565; 10,256; 11,328; 7,179	1	46,265,816.6	0.00	11,316	16,53,257
Case_25	0,8	2	12,565; 10,256	0.7	46,629,446.9	0.16	11,316	16,355,480
Case_26	0,8	2	12,565; 10,256	0.8	46,629,058.96	0.09	11,316	17,231,322
Case_27	0,8	2	12,565; 10,256	0.9	46,629,058.96	0.11	11,316	27,032,830
Case_28	0,8	2	12,565; 10,256	1	46,726,179.11	0.25	11,316	13,898,214
Case_29	0,8	3	12,565; 10,256; 11,328	0.7	46,358,752.74	0.00	11,316	6,578,495
Case_30	0,8	3	12,565; 10,256; 11,328	0.8	46,358,752.74	0.00	11,316	2,597,622
Case_31	0,8	3	12,565; 10,256; 11,328	0.9	46,358,752.74	0.00	11,316	4,195,500
Case_32	0,8	3	12,565; 10,256; 11,328	1	46,358,752.74	0.00	11,316	6,502,050
Case_33	0,8	4	12,565; 10,256; 11,328; 7,179	0.7	46,302,318.71	0.00	11,316	494,199
Case_34	0,8	4	12,565; 10,256; 11,328; 7,179	0.8	46,302,318.71	0.00	11,316	555,084
Case_35	0,8	4	12,565; 10,256; 11,328; 7,179	0.9	46,302,318.71	0.00	11,316	787,498
Case_36	0,8	4	12,565; 10,256; 11,328; 7,179	1	46,302,318.71	0.00	11,316	635,379

References

1. O'Kelly ME (1987) A quadratic integer program for the location of interacting hub facilities. Eur J Oper Res 32:393–404
2. Campbell JF (1994) A survey of network hub location. Stud Locat Anal 6:31–49
3. O'Kelly ME, Miller HJ (1994) The hub network design problem: a review and synthesis. J Transp Geogr 2:31–40
4. Skorin-Kapov D, Skorin-Kapov J, O'Kelly M (1996) Tight linear programming relaxations of uncapacitated p-hub median problems. Eur J Oper Res 73:501–508
5. Klincewicz JG (1991) Heuristics for the p-Hub location problem. Eur J Oper Res 79:25–37
6. Abdinnour-Helm S (1998) A hybrid heuristic for the uncapacitated hub location problem. Eur J Oper Res 106:489–499
7. Smith KA, Krishnamoorthy M, Palaniswami M (1996) Neural versus traditional approaches to the location of interacting hub facilities. Locat Sci 4:155–171
8. Hung HK, Rikkers RF (1974) A heuristic algorithm for the multi-period facility location problem, In: 45th joint national meeting of ORSA/TIMS, Boston
9. Rao RC, Rutenberg DP (1977) Multi location plant sizing and timing. Manag Sci 23:1187–1198
10. Luss H (1982) Operations research and capacity expansion problems: a survey. Oper Res 30:907–947
11. Cockburn A Dr (2008) Using both incremental and iterative development, STSC CrossTalk (USAF Software Technology Support Center) 21(5):27–30
12. Lehner F, Wildner S, Scholz S (2008) Wirtschaftsinformatik: Eine Einführung. München. Carl Hanser Verlag, München
13. Fotheringham AS (1983) A new set of spatial interaction models: the theory of competing destinations. Environ Plann 15–36
14. http://fa-technik.adfc.de/code/opengeodb/DE.tab. Effective: 13-02-14
15. AIMMS (2012) Paragon decision technology, http://www.aimms.com
16. CPLEX (2012) IBM Ilog, http://www.01.ibm.com/software/integration/optimization/cplex-optimizier/
17. Müller T, Hillebrandt J Paketproduktion (2012), Arbeitsbericht, Aachen, 26.04.2010
18. Campbell JF (2009) Hub location for time definite transportation. Comput Oper Res 36:3107–3116
19. Yaman H (2011) Allocation strategies in hub networks. Eur J Oper Res 211:442–451

Part V
Health Care Planning and Scheduling

A Mixed Integer Programming Approach to Surgery Scheduling with Simultaneous Decision Making

Halil Ibrahim Gündüz and Martin Nikolas Baumung

Abstract In recent years, hospitals have been affected by restrictive budgets that call for greater efficiency and a better resource utilization. A hospital's surgical suite is not only widely recognized as being one of the major cost drivers, but also has a huge impact on many other departments. It therefore is a priority to improve the efficiency of this particular component. This work covers a real case of surgeon and elective surgery scheduling for the Clinic Department of Otorhinolaryngology and Plastic Head and Neck Surgery in University Hospital Aachen, Germany, and aims at simplifying the regular planning process and improving the scheduling in order to reduce costs related to the operating room time while considering two types of resources—operating rooms and surgeons. Besides the scheduling of elective surgeries from a waiting list on a weekly horizon, the allocation of surgeries to operating rooms and the allocation of surgeons to the surgeries are also part of this planning process. For this purpose, we developed a mixed integer linear programming model, which aims at minimizing the costs for the operating room time required to perform all of the surgeries. Because of the clinic department's small size, and in contrast to comparable approaches that can be found in the literature, the proposed model addresses the scheduling and both of the allocation decisions simultaneously. The lower bound, which proved to be quite weak in the original model, could be improved drastically by adding valid inequalities. With this improved model, near optimal solutions can be computed within a couple of minutes, and an analysis revealed that these solutions comply with the conditions imposed by the hospital and provide a very good utilization of the operating rooms.

H.I. Gündüz (✉)
Deutsche Post Chair of Optimization of Distribution Networks,
RWTH Aachen University, Kackertstr. 7, 52072 Aachen, Germany
e-mail: guenduez@dpor.rwth-aachen.de

M.N. Baumung
Deutsche Post Chair of Optimization of Distribution Networks,
RWTH Aachen University, Kackertstr. 7, 52072 Aachen, Germany
e-mail: baumung@dpor.rwth-aachen.de

H.-J. Sebastian et al. (eds.), *Quantitative Approaches in Logistics and Supply Chain Management*, Lecture Notes in Logistics,
DOI 10.1007/978-3-319-12856-6_8

1 Introduction

In the last decades the health care sector in Germany has been affected by many budget cuts that call for greater efficiency in the use of available resources in hospitals. In particular the diagnostic—related group (DRG) was developed in the early 1980s as a patient classification system to replace the cost-based reimbursement that had been used up to that point. Patients within each group are regarded as clinically similar and are expected to use the same level of hospital resources. The purpose of the DRGs is to determine the payment of medicare to the hospital per group. While in most countries, hospital-based DRGs are used for distribution of the health insurance-related budget, in Germany they were transformed into a lump-sum system in 2003 and called the German-DRG (G-DRG). Since then it has been used for the settlement of prices for the various types of treatment of individual cases. In addition to the need for a greater efficiency on account of budget cuts or lump-sum systems, some countries-similarly to Germany—are faced with an ageing population. Therefore, the need for a better utilization of hospital resources is important and urgent e.g. operating rooms and surgeons.

The operation theater has a direct impact on many divisions and is the central engine of any hospital. The impact affects, for example, the necessary provision of anesthesiologists, surgical nurses, surgical operation assistants, surgical wards, and recovery units. Therefore, it is important to improve the efficiency of operation theaters of hospitals in general or at least of the clinic department units of a hospital. An improvement of the efficiency may lead to an increased number of performed surgeries and thus, to a reduction in surgery waiting lists. Moreover, it also may result in a better utilization of anesthesiologists and surgical nurses and hence lead to increased productivity.

The remainder of this article is organized as follows. Existing approaches and related problems from the literature are surveyed in Sect. 2. The article proceeds in Sect. 3 with a description of the simultaneous surgery and surgeon scheduling problem, followed by a mixed integer formulation of the problem in Sect. 4. To accelerate the mixed integer solvers handling instances of the model, we introduce some improvements in Sect. 5. The model is refined in Sect. 6 with respect to the availability of operating rooms and surgeons over the time horizon. In Sect. 7 we survey computational results on problem instances of different sizes and diversity of surgeries. Finally, we conclude in Sect. 8 and give a prospect for future work.

2 Survey of the Related Literature

In the literature related to operation room planning, researchers usually differentiate between strategic, tactical, and operational decisions. Strategic decisions, for instance, comprise case mix planning, defining a hospital's surgery supply on a long-term basis. Master surgery planning, defining operating room and surgeon availability, is an example of a tactical planning problem studied in the literature. In this paper we focus on surgery scheduling, which is an operational problem thoroughly studied. Scheduling problems can be categorized with respect to the classes of patients considered, namely elective and non-elective patients. For a comprehensive review of the related literature, we refer to [2, 6, 10, 13].

According to [2], surgery scheduling can be divided into two separate scheduling processes: *Advance scheduling*, which schedules each surgery for a specific day, and *allocation scheduling*, which determines the exact starting time and the operating room that the surgery is performed in for all surgeries allocated to the same day. Solving both scheduling problems together gives the best solutions in terms of quality and feasibility, since both problems interact with each other. As solving both problems simultaneously usually requires high computational resources, both types of scheduling are often studied separately in the literature.

Works which focus on *advance scheduling* only, include, among others [8, 11, 14]. Reference [8] introduce a column-generation based approach to allocate surgeries to operating rooms and days with the objective to minimize the cost of unexploited opening hours and overtime. In [11], the authors consider the robust surgery loading problem for a hospital's operating theater department and assign surgeries and sufficient planned slack to operating room days while maximizing capacity utilization and minimizing the risk of having to cancel surgeries. Reference [14] provide an integer linear problem used to assign elective surgeries to operating rooms while mastering the risk of no realization and stabilizing the operating rooms' utilization time.

A number of studies focus on the *allocation scheduling* problem only. References [4, 5] formulate a multiple objective optimization model for scheduling elective surgeries on a daily basis. Reference [1] investigate the impact of allowing patient recovery in the operating room when no recovery bed is available and propose a Lagrangian relaxation-based method to solve this particular operating theater scheduling problem. In [9] the authors provide a mixed integer linear problem used to determine the allocation of operating rooms and surgeons to surgeries, as well as the sequence of surgeries within the individual operating rooms. The authors consider various resources and operative constraints, but since the scheduling is done on a daily basis, the problem is still simple enough to be solved exactly. References [7, 12] both developed stochastic approaches for the *allocation scheduling* problem.

Some works consider *advance scheduling* and *allocation scheduling* in a single problem. Reference [17] propose a formulation that includes both the planning and scheduling of the surgeries and develop a heuristic procedure based on a genetic algorithm to solve the resulting hard optimization problem. However, the set of surgeries assigned to a particular surgeon is known in advance and is not determined by the algorithm. Reference [16] present a meta-heuristic algorithm for a model assigning operating rooms and dates to a set of elective surgeries, as well as scheduling the surgeries of each day and room and, simultaneously, creating a schedule for each surgeon. In [15], the authors formulate an integer linear programming model scheduling elective surgeries from a waiting list on a weekly basis while maximizing the use of the operating rooms. However, as in [17], the allocation of surgeries to surgeons as well as the sequence of surgeries for each surgeon is given in advance and not determined by the model. Non-optimal solutions are improved by a simple and efficient heuristic. Reference [3] suggest a mixed integer linear problem approach, which combines medium-term planning for surgery with short-term scheduling of resources. Surgeons are treated as a resource and the need for a surgery for a specific resource is given by parameters. The above mentioned research has much in common with the model proposed below, except for the allocation of surgeons to surgeries.

In this work, surgeries and surgeons are scheduled simultaneously. A full problem description is given in the next section.

3 Simultaneous Surgery and Surgeon Scheduling

This work is based on a cooperation with the Clinic Department of Otorhinolaryngology and Plastic Head and Neck Surgery incorporated in University Hospital Aachen and investigates the tactical problem of simultaneous surgery scheduling and the assignment of surgeons and operating rooms. With 34 specialist clinics, 25 institutes, and five interdisciplinary units, University Hospital Aachen covers the entire medical spectrum. It currently has around 1,240 beds and provides medical care for approximately 47,000 inpatients and 153,000 outpatients per year. Overall, 52 operating rooms are available in the operation theater for all clinics. Depending on the number of the urgency and surgery waiting list of each clinic, the operation management weekly assigns operating rooms to each clinic. In return, each clinic department has to submit a weekly time schedule of planned surgeries so that operations management can build a schedule for the necessary anesthesiologist, nursing teams, and mobile specialized equipment.

In this paper, we provide a model which enables the Clinic Department of Otorhinolaryngology and Plastic Head and Neck Surgery to establish a surgery schedule for the week to come based on the assigned operating rooms and the list of surgeries to be performed. At this point, we consider elective surgeries only, since for a weekly planning, only surgeries that can be well planned in advance can be taken into account adequately. The schedule for elective surgeries is constructed considering the following decisions:

1. Surgery assignment to days: Each surgery is assigned to a specific day within the planning horizon, i.e. the week ahead.
2. Surgery assignment to operating rooms: Each surgery is assigned to a specific operating room.
3. Surgery assignment to surgeons: Each surgery is assigned to a surgeon, who then performs the surgery.
4. Surgery sequencing: All surgeries that have been assigned to the same operating room on the same day are sequenced such that the starting and ending time of each surgery is determined.

It seemed appropriate to make all the above mentioned decisions simultaneously because this approach yields better solutions than a sequential one and because the clinic department is relatively small in size (e.g. 2 operating rooms and 4 surgeons and approximately 30 surgeries per week), what keeps the resulting mathematical problem tractable.

Other decisions, such as the allocation of nurses, anesthesiologists, etc., to surgeries are left aside, since they lie beyond the authority of the clinic department.

When planning a schedule, we pursue the goal of maximizing the utilization of the assigned operating rooms, which helps to increase a hospital's overall efficiency. The operating rooms are available from 8.00 am to 4.00 pm, from Monday to Friday, and need cleaning and disinfection, taking about 30 min, between each surgery.

4 Model Formulation

We give a formulation of the 5-day-ahead simultaneous surgery and surgeon scheduling problem as a linear mixed-integer programming problem. For an overview of the used symbols, parameters, and decision variables, we refer to Tables 1, 2 and 3. The set of the elective surgeries is denoted by I, the set of available surgeons by K, the set of available operating rooms by S, and the set of days of the time horizon by D (Mon., Tue., Wed., Thurs., and Fri.).

We use four classes of binary and three classes of continuous decision variables to formulate the model. The first decision of the model concerns the use of an operating room on a day d, for which we introduce the binary decision variables $y_{rd} \in \{0, 1\}$. If $y_{rd} = 1$ holds, then at least one surgery will take place in operation room r on day d. For the assignment of a surgery to a surgeon, an operating room, and a day, we introduce decision variables $x_{irkd} \in \{0, 1\}$. If $x_{irkd} = 1$ holds, then surgery i will be assigned to surgeon k and operation room r on day d. The next two classes of binary variables concern the precedence relations between each pair of surgeries i and j. The first one, $z_{ijrd} \in \{0, 1\}$, refers to precedence relation in an operating room r on day d and $z_{ijrd} = 1$ holds only if surgery i is performed (not necessarily directly)

Table 1 Symbols used in the model formulation

Symbols	Meaning
I	Set of all elective surgeries
K	Set of all available surgeons
R	Set of all available operating rooms
D	Set of days
i, j	Surgery $i, j \in I$
k	Surgeon $k \in K$
r, r_1, r_2	Operating room $r, r_1, r_2 \in R$
d, d_1, d_2	Day $d, d_1, d_2 \in D$

Table 2 Decision variables used in the model formulation

Variables	Meaning
$y_{rd} \in \{0, 1\}$	= 1 iff operating room r is used on day d
$x_{irkd} \in \{0, 1\}$	= 1 iff surgery i is performed by surgeon k on day d in operating room r
$z_{ijrd} \in \{0, 1\}$	= 1 iff surgery i is performed before j on day d in same operating room r
$u_{ijkd} \in \{0, 1\}$	= 1 iff surgery i is performed before j on day d by same surgeon k
$b_{irkd} \in \mathbb{R}_+$	Start time of surgery i performed by surgeon k on day d in operating room r
$f_{irkd} \in \mathbb{R}_+$	End time of surgery i performed by surgeon k on day d in operating room r
$t_{rd} \in \mathbb{R}_+$	End time of last surgery in operating room r on day d

Table 3 Parameters used in the model formulation

Parameters	Meaning
L_r^b	Last possible surgery start in operating room r
L_r^f	Last possible surgery end in operating room r
C_{rd}	Model fixed costs for using operating room r on day d
$\widetilde{C}_r$	Occupation costs per minute of operating room r
S^M	Modification time of an operating room between two consecutive surgeries
S^R	Switching time of a surgeon between different operating rooms
pre_i	Initiation time of surgery i
p_i	Duration of surgery i
$post_i$	Recovery from surgery i
$time_i$	$= pre_i + p_i + post_i$
M	Big M

before j in the same operating room and on the same day. Similarly, the second one, $u_{ijkd} \in \{0, 1\}$, refers to precedence relation of surgeon k on day d, and $u_{ijkd} = 1$ holds only if surgery i is performed (not necessarily directly) before j by the same surgeon and on the same day. In addition, we introduce variables concerning start times $b_{irkd} \in \mathbb{R}_+$ and end times $f_{irkd} \in \mathbb{R}_+$ for all surgeries i, surgeons k, operating rooms r, and days d. Note that $b_{irkd} = f_{irkd} = 0$ holds if one of the following restrictions is fulfilled:

- surgery i is not scheduled to operating room r
- surgery i is not assigned to surgeon k
- surgery i is not scheduled on day d

In our model, start and end time variables are coded as zero-based values. The value 0 represents the start time 8 a.m. and the start time of the first surgery of a day in an available operating room can have the value 0.

Finally, for the calculation of variable occupation costs we have to determine the end time of the last surgery on each day in each operating room. Therefore, we introduce the continuous decision variables $t_{rd} \in \mathbb{R}_+$.

Our objective is to find a cost optimal schedule for all surgeries. The most crucial cost component is the occupation costs of operating rooms. Of non importance for the underlying real costs are fixed costs for the use of an operating room. But we still use them to avoid symmetries in the solution space and to model schedules preferred by the surgeons and the management. Assume that an operating room is used on Monday and Tuesday. We obtain the same objective if we, for example, shift the Monday schedule to Wednesday and the Tuesday schedule to Friday. But the preferred solution of the management is to have schedules starting on Monday and continuing on the following days without a break. Let C_{rd} denote the model fixed costs for the use of operating room r on day d and let $\widetilde{C}_r$ denote the occupation costs per minute of operating room r. Then we have the following objective function:

$$\min z = \sum_{r,d} (C_{rd} y_{rd} + \widetilde{C}_r t_{rd}) \tag{1}$$

For the realization of the above mentioned preferred solution in our model, we use a strictly monotonically increasing fixed costs, i.e. $C_{rd_1} < C_{rd_2}$ for all $r \in R$ and $d_1, d_2 \in D$ with $d_1 < d_2$. Furthermore, sufficiently high fixed costs force the model to make an efficient use of the available surgery time per operating room.

The necessary constraints can be classified into three major groups: one for the assignment of surgeries, one for precedence relations between surgeries, and one for the (start and end) times of the surgeries. In our problem all surgeries must be scheduled during the regarded time horizon, i.e. each surgery must be exactly assigned to one surgeon, one operating room, and one day:

$$\sum_{r,k,d} x_{irkd} = 1 \quad \forall\, i \in I \tag{2}$$

Surgeries can only be assigned to an operating room r on day d if it is used on that day:

$$\sum_{i,k} x_{irkd} \leq M y_{rd} \quad \forall\, r \in R,\ d \in D \tag{3}$$

The next constraints concern the precedence relation between surgeries. In our sense, precedence relations between a pair of surgeries only occur if they are both performed in the same operation room and on the same day on the one hand and by the same surgeon and on the same day on the other hand. For the establishment of the precedences, we introduce the following three constraints:

$$z_{ijrd} + z_{jird} \leq \sum_{k} x_{irkd} \quad \forall\ i, j \in I, \quad i < j, \quad r \in R, \quad d \in D \tag{4}$$

If surgery i is not assigned to operating room r on day d, then the right-hand side of inequality (4) is 0. Thus, the left-hand side must also be 0, i.e. $z_{ijrd} = z_{jird} = 0$ holds. If surgery i is assigned to operating room r on day d, then it is also assigned exactly to one surgeon according to constraints (2), and the right-hand side of (4) equals 1. In this case surgery i can be in a precedence relation with j (predecessor or successor). Then only one of the binary variables z_{ijrd} and z_{jird} can equal to 1 but it is not mandatory:

$$z_{ijrd} + z_{jird} \leq \sum_{k} x_{jrkd} \quad \forall i, j \in I, \ i < j, \ r \in R, \ d \in D \tag{5}$$

Constraints (5) are equal to constraints (4) with switched roles of surgeries i and j:

$$z_{ijrd} + z_{jird} \geq \sum_{k} (x_{irkd} + x_{jrkd}) - 1 \quad \forall i, j \in I, \ i < j, \ r \in R, \ d \in D \tag{6}$$

Finally, if surgeries i and j are assigned to the same operating room r on same day d, then both are exactly assigned to one surgeon, and the right-hand side of inequality (6) is $2 - 1 = 1$. Then at least one of the binary variables z_{ijrd} and z_{jird} on the left-hand side must equal to 1 to fulfill the validity. In combination with constraints (4) and (5), only one of both variables can equal to 1. Thus, surgery i must be a predecessor or successor of j in the scheduling plan of operating room r on day d. In all other cases the right hand side equals to 0 or -1, and the inequality is automatically fulfilled because of the nonnegative values of all variables z_{ijrd}. In combination with constraints (4) and (5) it is obvious that then both variables z_{ijrd} and z_{jird} must equal to 0, and there is no precedence relation between surgeries i and j according to operating room r on day d.

By analogy with constraints (4)–(6), we introduce the following next three constraints concerning the precedence relations of surgeries according to surgeon and day assignment. The only difference is that the role of the operating room is replaced by a surgeon:

$$u_{ijkd} + u_{jikd} \leq \sum_{r} x_{irkd} \quad \forall i, j \in I, \ i < j, \ k \in K, \ d \in D \tag{7}$$

$$u_{ijrd} + u_{jird} \leq \sum_{r} x_{jrkd} \quad \forall i, j \in I, \ i < j, \ k \in K, \ d \in D \tag{8}$$

$$u_{ijkd} + u_{jikd} \geq \sum_{r} (x_{irkd} + x_{jrkd}) - 1 \quad \forall i, j \in I, \ i < j, \ k \in K, \ d \in D \tag{9}$$

Constraints (4)–(9) determine a sequence of scheduled surgeries per operating room, per surgeon, and per day, and therefore help to determine start and end times of each surgery and to avoid assignments of a surgeon to surgeries in different operating rooms on the same day with time overlap. If a surgery is not assigned to an operating room r and to surgeon k on day d, then the start and end time is set to 0 in our model by the following two constraints:

$$b_{irkd} \leq M x_{irkd} \quad \forall i \in I, \quad r \in R, \quad k \in K, \quad d \in D \tag{10}$$

$$f_{irkd} \leq Mx_{irkd} \quad \forall i \in I, \quad r \in R, \quad k \in K, \quad d \in D \tag{11}$$

Thus, in combination with constraints (2), variables b_{irkd} and f_{irkd} can have a positive nonzero value only for exactly one combination of $r \in R$, $k \in K$, and $d \in D$. To determine the end time of a surgery j, we take into account that it must be greater or equal to the maximum end time of all predecessors plus the overall duration and the modification time before the start of surgery j. Constraints (12) fulfill this purpose for the predecessors according to operating rooms and days:

$$\sum_{r,k} f_{jrkd} \geq \sum_{r,k} f_{irkd} + (time_j + S^M) - M(1 - \sum_r z_{ijrd})$$
$$\forall i, j \in I, i \neq j, \; r \in R, \; k \in K, \; d \in D \tag{12}$$

For a pair i and j of surgeries, the $\sum_r z_{ijrd}$ is equal to 0 if they are not in a precedence relation according to any operating room. Then the right hand side of constraints (12) is dominated by $-M$ and therefore is negative. Thus, the inequality is automatically satisfied. In the case where i is a predecessor of j, then $\sum_r z_{ijrd} = 1$ holds, and the inequality reduces itself to $\sum_{r,k} f_{jrkd} \geq \sum_{r,k} f_{irkd} + (time_j + S^M)$. Because of the precedence relation, both surgeries are assigned to the same operating room r_0 on the same day d_0. Further, both can be scheduled to different surgeons $k_1 \neq k_2$ or to the same surgeon $k_1 = k_2$. As mentioned above, only one of the variables f_{irkd} or f_{jrkd} can have a positive value. Then, the inequality reduces to $f_{jr_0k_1d_0} \geq f_{ir_0k_2d_0} + (time_j + S^M)$ for all pairs i, j, where i is a predecessor of j according to operating room r_0 and day d_0.

$$\sum_{r,k} f_{jrkd} \geq \sum_{r,k} f_{irkd} + (time_j + S^R) - M(1 - \sum_k u_{ijkd})$$
$$\forall i, j \in I, i \neq j, \; r \in R, \; k \in K, \; d \in D \tag{13}$$

By analogy with the same arguments, constraints (13) fulfill the above mentioned purpose for the predecessors according to surgeons and days. The role of operating rooms is replaced by surgeons and the modification time of operating rooms is replaced by the switching time of surgeons between two different operating rooms. Further, the latest end time of surgeries in operating rooms is restricted:

$$\sum_{r,k,d} f_{irkd} \leq L_r^f \quad \forall i \in I \tag{14}$$

For the efficient use of operating rooms, all assigned surgeries should end as early as possible, in particular the last surgery of each operating room and each day, i.e. unnecessary gaps between two consecutive surgeries should be avoided. Therefore, we introduce the following restrictions:

$$\sum_{k} f_{irkd} \leq t_{rd} \quad \forall\, i \in I,\ r \in R,\ d \in D \tag{15}$$

For a given operating room r_0 on day d_0, constraints (15) force $t_{r_0 d_0}$ to equal to the maximum of $f_{ir_0 kd_0}$. In combination with the objective (1), $t_{r_0 d_0}$ equals the maximum of $f_{ir_0 kd_0}$, which is the end time of the last surgery in operating room r_0 on day d_0. Thus, unnecessary gaps between two consecutive surgeries are forced to be eliminated by minimizing t_{rd}, i.e. the occupation costs. We determine the start times of surgeries by backward calculation:

$$b_{irkd} = f_{irkd} - x_{irkd}\, time_i \quad \forall\, i \in I,\ r \in R,\ k \in K,\ d \in D \tag{16}$$

If $x_{irkd} = 0$ holds, then $b_{irkd} = f_{irkd}$ applies in (16) and in addition, both variables are equal to 0 because of (10) and (11). In the case where $x_{irkd} = 1$ holds, then $b_{irkd} = f_{irkd} - time_i$ applies, i.e. the start time of a surgery is calculated by the end time minus the overall duration (including initiation and recovery). Furthermore, the switching time of a surgeon between two assigned consecutive surgeries in different operating rooms must be considered and linked to the end time and start time, respectively. A surgeon can leave surgeries with the start of the recovery phase and does not need to be present during the initiation phase:

$$\begin{aligned} &b_{jr_1kd} + pre_j \geq f_{ir_2kd} - post_i + S^R - M(1 - u_{ijkd}) \\ &\forall\, i, j \in I, i \neq j,\ r_1, r_2 \in R, r_1 \neq r_2,\ k \in K,\ d \in D \end{aligned} \tag{17}$$

Constraints (17) concern only surgeries assigned to different operating rooms and with existing precedence relation according to a surgeon. If there is no such precedence relation (i.e. $u_{ijkd} = 0$), the right-hand side of inequalities (17) is dominated by $-M$. Thus, the right-hand side is negative and the inequality is automatically satisfied. In all other cases the inequality reduces itself to $b_{jr_1kd} + pre_j \geq f_{ir_2kd} - post_i + S^R$, where $f_{ir_2kd} - post_i$ is the planned end time of the surgical procedure of i, and $b_{jr_1kd} + pre_j$ is the planned start time of the surgical procedure of j. Constraints (17) ensure that the planned end time plus the switching time between operating rooms is before the start time of surgical procedure for surgeries i and j, where j is the successor of i according to a surgeon and a day.

$$\sum_{r,k,d} b_{irkd} \leq L_r^b \quad \forall\, i \in I \tag{18}$$

Further, the last possible start time of surgeries in operating rooms is restricted by constraints (18).

5 Improving the Formulation

We describe a bin packing subproblem, symmetry breaking inequalities, and an estimation of costs to improve the formulation of the previous section. Our goal is to achieve a formulation that the used MILP solver is able to solve faster or at least to close the gap faster.

5.1 Subproblem: Bin Packing Problem

It is possible to determine the minimum of required operating rooms. If we only concentrate on the problem of assigning all surgeries to operating rooms and days, then the resulting problem is a bin packing problem, where the operating rooms per day correspond to a bin, the available time of operating rooms to the volume of the bins, the surgeries to the objects, and the overall duration of each surgery time to the volume of the corresponding object. Further, the available time of each operating room is the same, i.e. each bin has the volume but the problem can easily be extended to a heterogeneous problem. Our goal is to minimize the number of used bins such that all objects are uniquely assigned to a bin and volume restrictions of the used bins are fulfilled.

We use the binary decision variables $v_{rd} \in \{0, 1\}$. If $v_{rd} = 1$ holds, then the bin corresponding to operating room r and day d is used. For the assignment decision of surgery to a bin, we introduce decision variables $w_{ird} \in \{0, 1\}$. If $w_{ird} = 1$ holds, then surgery i is assigned to the bin corresponding to operation room r and day d. Note that the index number can be reduced by one. This can be achieved by enumerating all combinations of the pair (r, d) and mapping them to a single index. Then the classical model of a one-dimensional bin packing problem occurs. Nevertheless, we use the following model formulation to obtain a better overall cost lower bound for the MILP formulation in Sect. 4:

$$\min z = \sum_{r,d} v_{rd}$$

The objective function minimizes the number of used operating rooms over the time horizon D such that following constraints are satisfied:

$$\sum_{r,d} w_{ird} = 1 \quad \forall\, i \in I$$

Each surgery has to be uniquely assigned to a operating room and to a day:

$$\sum_{i} w_{ird}\, time_i + ((\sum w_{ird}) - 1)S^r \leq L^r\, y_{rd} \quad \forall\, r \in R,\ d \in D$$

The duration (including initiation and recovery) of the assigned surgeries plus the switching time between two surgeries must not exceed the available time of the operating rooms. Note that at the end of a day there is no switching time necessary, and therefore only $(\sum w_{ird}) - 1)$ switching times occur per operating room and day.

For the solution of the bin packing problem it is not important which of the operating rooms are used but for our main problem we prefer those with the earliest possible days of the time horizon, because of the strictly monotonically increasing fixed costs. Therefore, we add the following constraints, which are also known as symmetry breaking constraints:

$$v_{rd_1} \geq v_{rd_2} \quad \forall\, r \in R,\; d_1, d_2 \in D, d_1 < d_2$$

The objective value of this model provides a lower bound for the number of necessary operating rooms. For this purpose, we introduce a further parameter LB_r for the MILP model in Sect. 4, which represents the objective value of the bin packing subproblem, and we add constraints (19) to our MILP formulation:

$$\sum_{r,d} y_{rd} \geq LB_r \quad \forall\, r \in R,\; d \in D \tag{19}$$

Furthermore, we introduce parameter LB_{fix}. The calculation $LB_{fix} = \sum_{r,d} C_{rd}$ v_{rd} is a result of the bin packing problem, and we obtain a lower bound constraint for the overall fixed costs of our MILP formulation:

$$\sum_{r,d} C_{rd}\, y_{rd} \geq LB_{fix} \tag{20}$$

5.2 Symmetry Breaking Constraints

In Sect. 4 we use strictly monotonically increasing fixed costs to obtain a preferred solution of the management with schedules starting at the beginning of the time horizon and continuing on the following days without a break. But in the solution domain they are not forbidden. We can reduce the solution domain by adding the following constraints:

$$y_{rd_1} \geq y_{rd_2} \quad \forall\, r \in R,\; d_1, d_2 \in D, d_1 < d_2 \tag{21}$$

Constraints (21) allow the model only to use an operating room r on a day d_2 if the same operating room is used on all previous days d_1, i.e. $d_1 < d_2$. Note that feasible solutions of the MILP in Sect. 4 are eliminated by constraints (21) but because they are strictly monotonically increasing fixed costs, they are dominated by other feasible solutions. In the case of constant fixed costs, shifting the same schedules to different days will result in the same objective. With constraints (21) these symmetries are eliminated and therefore they are called symmetry breaking constraints.

Symmetries also occur through permutations of schedules according to the various available operating rooms on the same day. By analogy, we can add constraints (22) to break these symmetries, too:

$$y_{r_1 d} \geq y_{r_2 d} \quad \forall\, r_1, r_2 \in R, r_1 < r_2,\ d \in D \tag{22}$$

The symmetry breaking constraints can also be applied to the end time of each day's last surgeries:

$$t_{rd_1} \geq t_{rd_2} \quad \forall\, r \in R,\ d_1, d_2 \in D, d_1 < d_2 \tag{23}$$

$$t_{r_1 d} \geq t_{r_2 d} \quad \forall\, r_1, r_2 \in R, r_1 < r_2,\ d \in D \tag{24}$$

Note that constraints (21)–(24) can be applied in the described way as long as all operating rooms are available throughout the time horizon and have the same time capacity. Slight modifications are necessary if operating rooms are not available throughout the time horizon, and these will be presented in Sect. 6.

5.3 Further Estimations and Valid Inequalities

It is easy to estimate a lower bound for the overall required occupation time of operating rooms of the given set I of surgeries. First, we introduce the parameter $time_I = \sum_i time_i$, which corresponds to the sum of all surgery durations (including initiation and recovery). Further, we know that modification times occur after each surgery in each used operating room except the last surgery. In a worst case scenario, each operating room is used on every day, thus, overall $|R|$ times $|D|$. Then, $N = \min\{0, |I| - |R|\,|D|\}$ is the least number of times where modification times are taken into account:

$$\sum_{r,d} \widetilde{C}_r\, t_{rd} \geq \widetilde{C}_r(time_I + S^M(|I| - |N|)) \tag{25}$$

The right-hand side of constraint (25) includes the estimated lower bound of time for all surgery processes multiplied by the occupation costs per minute. Thus, the occupation costs of the model on the left-hand side must be greater or equal to them. A lifting is possible by replacing N with the to determined number of end-of-day surgeries:

$$\sum_{r,d} \widetilde{C}_r\, t_{rd} \geq \widetilde{C}_r(time_I + S^M(|I| - \sum_{r,d} y_{rd})) \tag{26}$$

The number of last surgeries equals to $\sum_{r,d} y_{rd}$, which is the overall number of used operating rooms over the time horizon. By aggregation of constraints (20) and (25) respectively (26), we obtain the following two constraints:

$$\sum_{r,d}(C_{rd}\ y_{rd} + \widetilde{C}_r\ t_{rd}) \geq LB_{fix} + \widetilde{C}_r(time_I + S^M(|I| - |N|)) \tag{27}$$

$$\sum_{r,d}(C_{rd}\ y_{rd} + \widetilde{C}_r\ t_{rd}) \geq LB_{fix} + \widetilde{C}_r(time_I + S^M(|I| - \sum_{r,d} y_{rd})) \tag{28}$$

The left-hand side of constraints (27) and (28) is the objective function from Sect. 4 and must meet the above described lower bound value. So far, constraints (14) represent time capacity constraints which can be strengthened. The time capacity of each operating room on each day is defined by the latest possible end of a surgery L_r^f. Further, time resources are used by surgery and modification durations. Constraints (29) restrict the time resource use by the given time capacity:

$$\sum_{i,k} x_{irkd}\ time_i + ((\sum_{i,k} x_{irkd}) - y_{rd})S^M \leq L_r^f\ y_{rd} \quad \forall\, r \in R,\ d \in D \tag{29}$$

If operating room r is used on day d, the right-hand side is L_r^f, otherwise 0, and none of the surgeries can be assigned to r on day d. The term $\sum_{i,k} x_{irkd}\ time_i$ is the duration of assigned surgeries to r on day d, and the term $(\sum_{i,k} x_{irkd}) - y_{rd}$ is the number of required modifications in r. Thus, the left-hand side represents the sum of surgery and modification durations in operating room r on day d. Note that gaps caused by surgeons' switching times between different operating rooms are not included.

$$\sum_{i,r,k,d} x_{irkd}\ time_i + ((\sum_{i,r,k,d} x_{irkd}) - \sum_{rd} y_{rd})S^M \leq L_r^f \sum_{rd} y_{rd} \tag{30}$$

Constraint (30) is the aggregate of constraints (29) over all operating rooms and the time horizon. On the right-hand side of both inequalities we take into account the latest possible end of a surgery L_r^f. But with variable t_{rd} we determine the end of the last surgery in operating room r on day d. Thus, we can replace $L_r^f\ y_{rd}$ by t_{rd} and get the constraints (31) and (32):

$$\sum_{i,k} x_{irkd}\ time_i + ((\sum_{i,k} x_{irkd}) - y_{rd})S^M \leq t_{rd} \quad \forall\, r \in R,\ d \in D \tag{31}$$

$$\sum_{i,r,k,d} x_{irkd}\ time_i + ((\sum_{i,r,k,d} x_{irkd}) - \sum_{rd} y_{rd})S^M \leq \sum_{rd} t_{rd} \tag{32}$$

We observed commercial solvers generating first initial solutions with hardly unbounded t_{rd}. Because of the objective function in an optimal solution, t_{rd} should have the value 0 for not used operating rooms and at most the value L_r^f for used operating rooms. But these are implications of the overall model and are not yet included in our models' constraints. Therefore, we add the following constraints to the model:

$$t_{rd} \leq L_r^f \; y_{rd} \quad \forall\, r \in R,\; d \in D \tag{33}$$

If $y_{rd} = 0$ holds, then t_{rd} has to be set to 0, too. In the case where $y_{rd} = 1$ holds, then t_{rd} is restricted to the latest possible end of the surgeries.

6 Refining the Model

In our MILP formulation in Sect. 4 we assume that every surgeon and operating room is available throughout the time horizon. In reality, sometimes operating rooms and surgeons are not available. In order to consider this, we introduce parameters $A_{rd} \in \{0, 1\}$ and $A_{kd} \in \{0, 1\}$. If $A_{rd} = 1$ holds, then operating room r is available on day d, otherwise $A_{rd} = 0$ holds. The same applies for A_{kd} with surgeon k and day d. The MILP formulation must be extended with the following constraints:

$$y_{rd} \leq A_{rd} \quad \forall\, r \in R,\; d \in D \tag{34}$$

It is obvious that if $A_{rd} = 0$ holds, then y_{rd} is also 0, and therefore operating room r on day d cannot be used.

$$\sum_{i,r} x_{irkd} \leq M\; A_{kd} \quad \forall\, r \in R,\; d \in D \tag{35}$$

If $A_{kd} = 1$ applies then the right-hand side of (35) has a big value and therefore it is valid. For the other case, where $A_{kd} = 0$ applies, the right-hand side equals to 0. Then, none of the surgeries and operating rooms can be assigned to surgeon k on day d. Thus, $\sum_{i,r} x_{irkd} = 0$ must hold to fulfill the inequality.

With the given non availability information we can also perform a short preprocessing by fixing the following variables to the value 0. We fix variables y_{rd} and t_{rd} for $r \in R$ and $d \in D$ with $A_{rd} = 0$. Variables x_{irkd}, b_{irkd}, and f_{irkd} are also fixed if $A_{kd} = 0$ or $A_{rd} = 0$ holds. We fix precedence relation variables z_{ijrd} if $A_{rd} = 0$ holds. Moreover, variables u_{ijkd} are fixed if $A_{kd} = 0$ holds.

In some cases, e.g. medical reasons, surgeries must be assigned to a particular surgeon (e.g. a professor or at least an assistant medical director). For this purpose, we introduce a parameter $A_{ik} \in \{0, 1\}$ and the corresponding necessary constraints (36):

$$\sum_{r,d} x_{irkd} \geq A_{ik} \quad \forall\, i \in I,\; k \in K \tag{36}$$

For surgeries which do not need to be assigned to a particular surgeon, the parameter A_{ik} has the value 0. Then, the inequality is automatically satisfied. On the other hand, if at least one parameter A_{ik} has the value 1, then surgeon k must be assigned to surgery i in any of the operating rooms and on any day of the time horizon. In combination with constraints (2), then $\sum_{r,d} x_{irkd} = 1$ must hold.

In addition, the symmetry breaking (21) and (22) constraints must be slightly modified to suit these requirements:

$$y_{rd_1} \geq y_{rd_2} \quad \forall\, r \in R,\; d_1, d_2 \in D, d_1 < d_2 \text{ and } A_{rd_1} = 1$$

$$y_{r_1d} \geq y_{r_2d} \quad \forall\, r_1, r_2 \in R, r_1 < r_2,\; d \in D \text{ and } A_{r_1d} = 1$$

By analogy, the same applies for constraints (23) and (24):

$$t_{rd_1} \geq t_{rd_2} \quad \forall\, r \in R,\; d_1, d_2 \in D, d_1 < d_2 \text{ and } A_{rd_1} = 1$$

$$t_{r_1d} \geq t_{r_2d} \quad \forall\, r_1, r_2 \in R, r_1 < r_2,\; d \in D \text{ and } A_{r_1d} = 1$$

Parameters A_{rd} and A_{kd} with value 1 indicate that operating room r and surgeon k are available on day d but in some cases they are only partially available during the day. To catch these cases, further parameters for time windows $[a, b]$ are necessary. We introduce L^a_{rd} and L^a_{kd} for the earliest availability time of operating room r and surgeon k on day d, respectively. L^b_{rd} and L^b_{kd} denote the latest availability time of operating room r and surgeon k on day d, respectively. Multiple time windows could be applied by virtual duplication of the operating room r and surgeon k, respectively. In the case of applied time windows, the volumes of the bin of the introduced bin packing problem in Sect. 5.1 must be adjusted to a heterogenous version. Time windows for surgeries are sometimes reasonable and necessary, too. For example, children's surgeries should be in the morning because children are usually impatient and cannot go for as long as adults without eating or drinking. In the same way, L^a_i and L^b_i describe the time window during which surgery i must be performed:

$$L^a_{rd}\Big(\sum_k x_{irkd}\Big) \leq \sum_k b_{irkd} \quad \forall\, i \in I,\; r \in R,\; d \in D, \text{ and } A_{rd} = 1 \tag{37}$$

$$L^b_{rd}\Big(\sum_k x_{irkd}\Big) \geq \sum_k f_{irkd} \quad \forall\, i \in I,\; r \in R,\; d \in D, \text{ and } A_{rd} = 1 \tag{38}$$

In the case where a surgery i_0 is assigned to operating room r_0 on day d_0, then the expressions $b_{i_0r_0kd_0}$ and $f_{i_0r_0kd_0}$ have one nonzero value (exception: $b_{i_0r_0kd_0}$ can have the value 0 if i_0 is the first surgery of the day), say for example k_0 ($x_{i_0r_0k_0d_0} = 1$). Then, constraints (37) and (38) are reduced to $L^a_{r_0d_0} \leq b_{i_0r_0k_0d_0}$ and $L^b_{r_0d_0} \geq f_{i_0r_0k_0d_0}$, respectively. Thus, constraints (37) and (38) force a surgery's start time to be after the earliest availability time and the end time to be before the latest possible availability time of operating room on a certain day. If a surgery is not assigned to an operating room on a certain day, then the right- and left-hand sides of constraints (37) and (38) have the same value, namely 0, and are therefore satisfied.

$$L_{kd}^{a}(\sum_{r} x_{irkd}) \leq \sum_{r} b_{irkd} \quad \forall\, i \in I,\ k \in K,\ d \in D, \text{ and } A_{rd} = 1 \tag{39}$$

$$L_{kd}^{b}(\sum_{r} x_{irkd}) \geq \sum_{r} f_{irkd} \quad \forall\, i \in I,\ k \in K,\ d \in D, \text{ and } A_{rd} = 1 \tag{40}$$

The same arguments apply for (39) and (40) with reverse role plays of surgeons and operating rooms. They both enforce that surgeries assigned to a certain surgeon must be scheduled during the availability time window of that surgeon.

$$L_{i}^{a} \leq \sum_{r,k,d} b_{irkd} \quad \forall\, i \in I \tag{41}$$

$$L_{i}^{b} \geq \sum_{r,k,d} f_{irkd} \quad \forall\, i \in I \tag{42}$$

Using the same arguments according to b_{irkd} and f_{irkd} outlined above, constraints (41) and (42) force the start and end time of each surgery to be within the intended time window.

Further, some surgeries include an infectious patient, and cleaning/disinfection takes much longer than the general switching time. It is therefore necessary to schedule these surgeries to the end of a day. The cleaning/disinfection can also take place after the last possible end of a surgery, and the department does not then have to pay for it. To distinguish between infectious and non-infectious patients, we require a further parameter $p_i \in \{0, 1\}$, where the value 1 represents an infectious patient at surgery i:

$$p_i \sum_{j} z_{ijrd} \leq M(1 - \sum_{k} x_{irkd}) \quad \forall\, i \in I,\ r \in R,\ d \in D \tag{43}$$

The inequality is always met if the patient of surgery i is not infectious or if surgery i is not assigned to operating room r on day d. Otherwise, the inequality is reduced to $\sum_j z_{ijrd} \leq 0$ for a pair (r_0, d_0) and thus, with nonnegative variables, $\sum_j z_{ijr_0d_0} = 0$ applies. This means that surgery i has no successors in the assigned operating room r_0 on day d_0 and constraints (37) serve their purpose. Note that for a feasible solution, $\sum_i p_i \leq \sum_{r,d} A_{rd}$ must hold. The problem of too many infectious patients can be avoided during the selection of the elective surgeries.

7 Computational Results

In this section, the mathematical model presented in Sect. 4, including the valid inequalities described there, is tested with different data sets. The mixed-integer problem was implemented in AIMMS 3.13 $\times$ 64 associated with the solver CPLEX 12.5 $\times$ 64. The maximum computation time for the solution process was limited to 7,200 seconds and a single core per instance on a PC running under Windows® 7 and equipped with a 3.4GHz Intel© Core™i7 processor.

7.1 Data Sets

The model was evaluated by solving a total of 120 test instances, where every instance consists of a set of operating rooms, a set of surgeons, and a set of surgeries that are to be performed, including information on the initiation time, the duration, and the recovery time for each surgery. The instances are divided into the two categories small (S) and large (L). The small instances feature two operating rooms, four surgeons, and a total of thirty surgeries, and therefore correspond to the size of the Clinic Department of Otorhinolaryngology and Plastic Head and Neck Surgery, whereas the large instances consider four operating rooms, eight surgeons, and sixty surgeries. Furthermore, we distinguish between low (L) and high (H) heterogeneity of the surgeries' durations. Eventually, we also distinguish between three different levels of total operating room usage. In instances with a low (L) usage level, on average 50 % of the operating rooms are required, whereas 75 and 100 % are required for the instances with medium (M) and high (H) usage level, respectively. All three characteristics can be combined to describe an instance; therefore LHM describes a large instance with a high duration volatility and a medium capacity usage. Ten different instances have been generated for every of the twelve possible combinations of the above mentioned characteristics, resulting in a total of 120 instances.

7.2 Results

Table 4 shows the results obtained for the small test instances, featuring two operating rooms, four surgeons, and thirty surgeries, and therefore comparable in size to the Clinic Department of Otorhinolaryngology and Plastic Head and Neck Surgery. With exception of the instances 02SHH and 09SHH, all instances were solved to optimality within the two hours of granted computational time, many of them within a couple of minutes. Even though the instances 02SHH and 09SHH could not be solved to optimality, the solutions found are still very good with gaps of 0.1 and 0.02 percent, respectively.

Table 4 Results for the small test instances with a low, medium, and high usage level

Instance	Solution	CPU time (s)	Gap	Instance	Solution	CPU time (s)	Gap	Instance	Solution	CPU time (s)	Gap
01SLL	96,304	575.7	Optimal	01SLM	158,804	154.9	Optimal	01SLH	221,304	2,936.1	Optimal
02SLL	96,240	662.7	Optimal	02SLM	158,708	303.7	Optimal	02SLH	221,176	625.8	Optimal
03SLL	122,688	158.7	Optimal	03SLM	159,572	49.9	Optimal	03SLH	222,328	666.9	Optimal
04SLL	103,316	59.1	Optimal	04SLM	159,572	112.0	Optimal	04SLH	222,328	207.8	Optimal
05SLL	122,144	208.7	Optimal	05SLM	158,756	74.7	Optimal	05SLH	221,240	369.4	Optimal
06SLL	96,656	797.1	Optimal	06SLM	159,332	159.2	Optimal	06SLH	222,008	3,349.9	Optimal
07SLL	96,656	970.7	Optimal	07SLM	159,332	290.6	Optimal	07SLH	222,008	664.3	Optimal
08SLL	103,092	120.0	Optimal	08SLM	159,236	97.9	Optimal	08SLH	221,880	654.2	Optimal
09SLL	102,868	138.7	Optimal	09SLM	158,900	361.2	Optimal	09SLH	221,432	672.4	Optimal
10SLL	96,432	1,025.5	Optimal	10SLM	158,996	229.0	Optimal	10SLH	221,560	498.3	Optimal
01SHL	122,304	154.1	Optimal	01SHM	158,996	222.4	Optimal	01SHH	189,188	770.5	Optimal
02SHL	136,072	139.0	Optimal	02SHM	160,148	174.2	Optimal	02SHH	190,916	7,200	0.10 %
03SHL	103,188	665.2	Optimal	03SHM	159,380	45.0	Optimal	03SHH	189.700	1,678.4	Optimal
04SHL	122,432	423.9	Optimal	04SHM	159,188	338.8	Optimal	04SHH	189.444	567.2	Optimal
05SHL	96,336	407.8	Optimal	05SHM	158,852	92.2	Optimal	05SHH	188.996	217.5	Optimal
06SHL	122,816	73.9	Optimal	06SHM	159,764	191.4	Optimal	06SHH	190.212	6,867.9	Optimal
07SHL	122,656	157.8	Optimal	07SHM	159,524	204.2	Optimal	07SHH	189.892	560.1	Optimal
08SHL	103,188	850.4	Optimal	08SHM	159,380	301.5	Optimal	08SHH	189.700	2,563.3	Optimal
09SHL	123,040	118.7	Optimal	09SHM	160,100	62.2	Optimal	09SHH	190.692	7,200	0.02 %
10SHL	101,972	184.5	Optimal	10SHM	138,184	1,139.3	Optimal	10SHH	154,896	227.6	Optimal

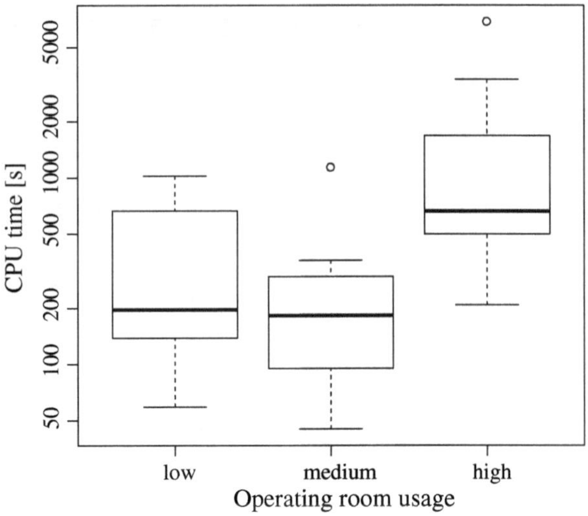

Fig. 1 Boxplot of computation time with respect to operating room usage

As one can see from Table 4, the computation time required to solve the instances varies greatly. A logarithmic boxplot of the CPU times required to solve the different instances to optimality in Fig. 1 also reveals that, on average, the computation time is much higher for the instances featuring a high operating room usage. This can be explained by the fact that, when operating room usage is close to 100 %, finding a feasible solution becomes increasingly difficult. However, we still find optimal solutions to all but two instances.

7.3 Valid Inequalities

To study the effectiveness of the valid inequalities, we also implemented the model introduced in Sect. 4 without the valid inequalities and symmetry breaking constraints described in Sect. 5 and compared the results with the ones obtained for the small sized instances from Sect. 7.2.

Table 5 gives the results for the small test instances, where gap 1 is the gap provided by the solver and gap 2 is the gap with respect to the optimal solutions given in Sect. 7.2. As we can see, no instance was solved to optimality within 2 hours of computation time and no feasible solution was found for any of the instances with a high operating room usage. The average optimality gap for the instances where a solution was found is around 36 %. We also notice that, where applicable, gap 1 is much larger than gap2, meaning that the lower bound provided by the model without the valid inequalities is very poor. Therefore, the formulation featuring the valid inequalities is clearly superior.

Table 5 Results for the small test instances with a low, medium, and high usage level without the valid inequalities

Instance	Solution	CPU time (s)	Gap 1(%)	Gap 2(%)	Instance	Solution	CPU time (s)	Gap 1(%)	Gap 2(%)	Instance	Solution	CPU time (s)	Gap 1	Gap 2
01SLL	135,176	7,200	1,665.99	40.36	01SLM	185,940	7,200	2,223.09	17.09	01SLH	n.a.	7,200	n.a.	n.a.
02SLL	154,644	7,200	1,937.61	60.69	02SLM	176,396	7,200	2,135.11	11.14	02SLH	n.a.	7,200	n.a.	n.a.
03SLL	144,780	7,200	1,792.05	18.01	03SLM	n.a.	7,200	n.a.	n.a.	03SLH	n.a.	7,200	n.a.	n.a.
04SLL	202,448	7,200	2,556.13	95.95	04SLM	208,912	7,200	2,510.09	30.92	04SLH	n.a.	7,200	n.a.	n.a.
05SLL	136,712	7,200	1,686.39	11.93	05SLM	184,820	7,200	2,270.22	16.42	05SLH	n.a.	7,200	n.a.	n.a.
06SLL	180,772	7,200	2,251.05	87.03	06SLM	218,296	7,200	2,646.67	37.01	06SLH	n.a.	7,200	n.a.	n.a.
07SLL	151,088	7,200	1,890.60	56.32	07SLM	185,780	7,200	2,211.85	16.60	07SLH	n.a.	7,200	n.a.	n.a.
08SLL	175,804	7,200	2,216.28	70.53	08SLM	206,432	7,200	2,499.28	29.64	08SLH	n.a.	7,200	n.a.	n.a.
09SLL	143,052	7,200	1,799.89	39.06	09SLM	218,312	7,200	2,627.54	37.39	09SLH	n.a.	7,200	n.a.	n.a.
10SLL	169,596	7,200	2,134.22	75.87	10SLM	163,028	7,200	1,976.80	2.54	10SLH	n.a.	7,200	n.a.	n.a.
01SHL	144,492	7,200	1,795.18	18.14	01SHM	213,796	7,200	2,727.04	34.47	01SHH	n.a.	7,200	n.a.	n.a.
02SHL	213,944	7,200	2,637.83	57.23	02SHM	214,724	7,200	2,603.13	34.08	02SHH	n.a.	7,200	n.a.	n.a.
03SHL	144,108	7,200	1,728.30	39.66	03SHM	194,360	7,200	2,347.68	21.95	03SHH	n.a.	7,200	n.a.	n.a.
04SHL	158,260	7,200	1,942.59	29.26	04SHM	217,464	7,200	2,688.28	36.61	04SHH	n.a.	7,200	n.a.	n.a.
05SHL	130,852	7,200	1,653.39	35.83	05SHM	215,908	7,200	2,701.54	35.92	05SHH	n.a.	7,200	n.a.	n.a.
06SHL	190,344	7,200	2,391.88	54.98	06SHM	214,932	7,200	2,724.64	34.53	06SHH	n.a.	7,200	n.a.	n.a.
07SHL	135,880	7,200	1,646.02	10.78	07SHM	183,536	7,200	2,271.41	15.05	07SHH	n.a.	7,200	n.a.	n.a.
08SHL	156,436	7,200	1,935.35	51.60	08SHM	214,756	7,200	2,686.85	34.74	08SHH	n.a.	7,200	n.a.	n.a.
09SHL	155,248	7,200	1,944.70	26.18	09SHM	186,964	7,200	2,172.05	16.78	09SHH	n.a.	7,200	n.a.	n.a.
10SHL	155,700	7,200	2,014.00	52.69	10SHM	185,860	7,200	2,253.88	34.50	10SHH	n.a.	7,200	n.a.	n.a.

Table 6 Results for the large test instances with a low, medium, and high usage level

Instance	Solution	CPU time [S]	Gap	Instance	Solution	CPU time [s]	Gap	Instance	Solution	CPU time [s]	Gap
01LLL	N.a.	7,200	N.a.	01LLM	N.a.	7,200	N.a.	01LLH	N.a.	7,200	N.a.
02LLL	N.a.	7,200	N.a.	02LLM	N.a.	7,200	N.a.	02LLH	412,060	7,200	0.29 %
03LLL	N.a.	7,200	N.a.	03LLM	N.a.	7,200	N.a.	03LLH	446,640	7,200	8.48 %
04LLL	N.a.	7,200	N.a.	04LLM	N.a.	7,200	N.a.	04LLH	N.a.	7,200	N.a.
05LLL	N.a.	7,200	N.a.	05LLM	N.a.	7,200	N.a.	05LLH	N.a.	7,200	N.a.
06LLL	N.a.	7,200	N.a.	06LLM	N.a.	7,200	N.a.	06LLH	N.a.	7,200	N.a.
07LLL	N.a.	7,200	N.a.	07LLM	N.a.	7,200	N.a.	07LLH	N.a.	7,200	N.a.
08LLL	N.a.	7,200	N.a.	08LLM	N.a.	7,200	N.a.	08LLH	416,156	7,200	1.30 %
09LLL	N.a.	7,200	N.a.	09LLM	N.a.	7,200	N.a.	09LLH	N.a.	7,200	N.a.
10LLL	N.a.	7,200	N.a.	10LLM	N.a.	7,200	N.a.	10LLH	N.a.	7,200	N.a.
01LHL	N.a.	7,200	N.a.	01LHM	N.a.	7,200	N.a.	01LHH	N.a.	7,200	N.a.
02LHL	248,640	7,200	1.32 %	02LHM	N.a.	7,200	N.a.	02LHH	N.a.	7,200	N.a.
03LHL	N.a.	7,200	N.a.	03LHM	N.a.	7,200	N.a.	03LHH	N.a.	7,200	N.a.
04LHL	N.a.	7,200	N.a.	04LHM	N.a.	7,200	N.a.	04LHH	N.a.	7,200	N.a.
05LHL	N.a.	7,200	N.a.	05LHM	N.a.	7,200	N.a.	05LHH	N.a.	7,200	N.a.
06LHL	N.a.	7,200	N.a.	06LHM	N.a.	7,200	N.a.	06LHH	N.a.	7,200	N.a.
07LHL	N.a.	7,200	N.a.	07LHM	N.a.	7,200	N.a.	07LHH	386,796	7,200	0.32 %
08LHL	N.a.	7,200	N.a.	08LHM	N.a.	7,200	N.a.	08LHH	N.a.	7,200	N.a.
09LHL	N.a.	7,200	N.a.	09LHM	N.a.	7,200	N.a.	09LHH	N.a.	7,200	N.a.
10LHL	N.a.	7,200	N.a.	10LHM	N.a.	7,200	N.a.	10LHH	N.a.	7,200	N.a.

7.4 Results for Large Instances

Even though this was not the primary target of this work, we also investigated the model's ability to handle instances corresponding to larger clinic departments. When looking at larger instances with four operating rooms, eight surgeons, and sixty surgeries, the model fails to find feasible solutions within the given computational time for almost all instances (see Table 6). This is mainly due to the sharply increasing problem size with a total of 279,640 variables, 222,020 of them being integer, and 2,122,411 constraints compared to 34,520 variables (27,310 of them being integer), and 91,626 constraints for the small instances. These results show that the model performs very well for small instances comparable in size to the environment found in the Clinic Department of Otorhinolaryngology and Plastic Head and Neck Surgery at University Hospital Aachen, but is clearly not suited for solving larger problems.

8 Conclusion and Future Work

In this paper, we address the problem of scheduling elective surgeries with simultaneous surgeon and operating room scheduling. We formulated a mixed integer problem to assign a set of elective surgeries to operating rooms and surgeons while sequencing them simultaneously. Taking into account many different real constraints in University Hospital Aachen, some special features of the proposed model were formed.

The proposed model was evaluated by solving several artificial test instances. Numerical results indicated that the model works very well for environments comparable in size to the Clinic Department of Otorhinolaryngology and Plastic Head and Neck Surgery. Almost all corresponding test instances were solved to optimality, and the gap evaluations for the instances not solved to optimality showed that the results were generally very good.

Because of the chosen formulation, the proposed model however is not apt to solve larger instances, representing environments such as they can be found in other clinic departments or other hospitals. Therefore, in future work, we will consider other model formulations and solution techniques to solve the described problem. A first simple heuristic, using the result of the solved bin packing problem to find an allocation of surgeries to operating rooms and days as an initial solution, shows some promising computational results for the large test instances.

Also, we are now working in a deterministic context. In future work, we will consider uncertainty related to surgery duration and the arrival of non-elective patients, in order to assess risks of overtime and surgery cancelation and possible operating room idle time resulting from the deterministic model for the scheduling problem that we considered here.

References

1. Augusto V, Xie X, Perdomo V (2010) Operating theatre scheduling with patient recovery in both operating rooms and recovery beds. Comput Ind Eng 58(2):231–238
2. Blake JT, Carter MW (1997) Surgical process scheduling: a structured review. J Soc Health Syst 5(3):17–30
3. Bulgarini N, Lorenzo D, Lori A, Matarrese D, Schoen F (2014) Operating room joint planning and scheduling. In: Matta A, Li J, Sahin E, Lanzarone E, Fowler J (eds) Proceedings of the international conference on health care systems engineering, Springer proceedings in mathematics and statistics, vol 61. Springer International Publishing, pp 127–138
4. Cardoen B, Demeulemeester E, Beliën J (2009) Optimizing a multiple objective surgical case sequencing problem. Int J Prod Econ 119(2):354–366
5. Cardoen B, Demeulemeester E, Beliën J (2009) Sequencing surgical cases in a day-care environment: an exact branch-and-price approach. Comput Oper Res 36(9):2660–2669
6. Cardoen B, Demeulemeester E, Beliën J (2010) Operating room planning and scheduling: a literature review. Eur J Oper Res 201(3):921–932
7. Denton B, Viapiano J, Vogl A (2007) Optimization of surgery sequencing and scheduling decisions under uncertainty. Health Care Manag Sci 10(1):13–24
8. Fei H, Chu C, Meskens N (2009) Solving a tactical operating room planning problem by a column-generation-based heuristic procedure with four criteria. Ann Oper Res 166(1):91–108
9. Ghazalbash S, Sepehri MM, Shadpour P, Atighehchian A (2012) Operating room scheduling in teaching hospitals. Adv Oper Res 2012:1–16
10. Guerriero F, Guido R (2011) Operational research in the management of the operating theatre: a survey. Health Care Manag Sci 14(1):89–114
11. Hans E, Wullink G, van Houdenhoven M, Kazemier G (2008) Robust surgery loading. Eur J Oper Res 185(3):1038–1050
12. Lamiri M, Grimaud F, Xie X (2009) Optimization methods for a stochastic surgery planning problem. Int J Prod Econ 120(2):400–410
13. Magerlein JM, Martin JB (1978) Surgical demand scheduling: a review. Health Serv Res 13(4):418–433
14. Marcon E, Kharraja S, Simonnet G (2003) The operating theatre planning by the follow-up of the risk of no realization. Int J Prod Econ 85(1):83–90
15. Marques I, Captivo ME (2012) An integer programming approach to elective surgery scheduling. OR Spectr 34(2):407–427
16. Riise A, Burke E (2011) Local search for the surgery admission planning problem. J Heuristics 17(4):389–414
17. Roland B, Martinelly CD, Riane F, Pochet Y (2010) Scheduling an operating theatre under human resource constraints. Comput Ind Eng 58(2):212–220